OPPORTUNITIES
FOR BIOMASS AND
ORGANIC WASTE VALORISATION

OPPORTUNITIES FOR BIOMASS AND ORGANIC WASTE VALORISATION

FINDING ALTERNATIVE SOLUTIONS TO DISPOSAL IN SOUTH AFRICA

Editors

Linda Godfrey

Johann F Görgens

Henry Roman

University of South Africa

Pretoria

Taylor & Francis Group

LONDON AND NEW YORK

© 2018 University of South Africa
First edition, first impression

Print book: ISBN 978-1-77615-010-6

Published by Unisa Press
University of South Africa
PO Box 392, 0003, UNISA

LONDON AND NEW YORK

Hardback ISBN13 978-0-367-19376-8
Paperback ISBN13 978-0-367-48885-7
Ebook ISBN13 978-0-429-20199-8

Prior to acceptance for publication by Unisa Press, this work was subjected to a double-blind peer review process mediated through the Senate Publications Committee of the University of South Africa.

Project Editors: Hetta Pieterse and Ingrid Stegmann
Book Designer, Typesetting and Cover Design: Monica Martins-Schuld
Copyeditor: Elize Zywotkiewicz
Indexer: Gail Malcolmson

Tel: 086 12 DALRO (from within South Africa), +27 (0)11 712-8000 Fax: +27 (0)11 403 9094 PO Box 31627, Braamfontein, 2017, South Africa
www.dalro.co.za

Contents

Quantifying organic waste in South Africa

Biological treatment

Mechanical and chemical treatment

Thermal treatment

Preface

The waste research, development and innovation (RDI) capability mapping, conducted by the Council for Scientific and Industrial Research (CSIR) and the Department of Science and Technology (DST) in 2014, showed that while waste RDI capability is generally still considered as "emerging", pockets of mature research excellence exist in certain South African universities and science councils. This provides a solid foundation for developing and strengthening research and development (R&D), human capital development (HCD) and innovation in the waste sector. However, much of the knowledge being produced by academia, remains locked up in post-graduate dissertations, and is often not made available to stakeholders in industry and government. The DST is exploring ways of unlocking this knowledge, in support of evidence-based decision-making. This book, the first in a series of planned academic publications, is aimed at bringing South Africa's research into the public domain, as part of the implementation of South Africa's 10-year Waste RDI Roadmap (DST, 2014).

Through a process of sector consultation, the Waste RDI Roadmap identified organic waste as one of five priority waste streams, requiring increased investment and support for R&D, HCD and Innovation. This book is intended to showcase South Africa's research on biomass and organic waste valorisation, as an alternative solution to disposal to landfill in South Africa. Organic waste is identified in the National Environmental Management: Waste Act 59 of 2008: National Waste Information Regulations, as one of 17 categories of general waste, consisting of garden, food and wood waste. Organic waste and commercially exploitable biomass resources (including forest biomass, sawmill biomass, sugar cane biomass and abattoir waste) are identified by the Department of Environmental Affairs (DEA) in the National Waste Information Baseline Report as the single largest general waste stream produced in South Africa. In 2011, South Africa generated in excess of 39 million tonnes of biomass and organic waste (DEA, 2012).

The terms *"organic waste"*, *"biomass waste"* (DST, 2013), *"biomass-based waste"*, *"biowastes"*, *"biomass-derived waste"*, *"waste agricultural biomass"*, *"agricultural or forestry residue"* and *"industrial co-products"* are often used interchangeably to reflect the biodegradable fraction of waste originating from the (i) agriculture, horticulture, aquaculture, forestry, hunting and fishing, food preparation and processing; (ii) wood processing; and (iii) domestic waste sectors. The United Nations Environment Programme (UNEP 2009:6) defines *"biomass waste"* as including *"agricultural wastes, such as corn stalks, straw, sugar cane leavings, bagasse, nutshells, and manure from cattle, poultry, and hogs; forestry residues, such as wood chips, bark, sawdust, timber slash, and mill scrap; municipal waste, such as waste paper and yard clippings."* The Japan Institute of Energy (JIE 2008:6) defines *"biomass"* as encompassing *"a wide variety including not only agricultural crops, timber, marine plants, and other conventional agriculture, forestry, and fisheries resources, but also pulp sludge, black liquor, alcohol fermentation stillage, and other organic industrial waste, municipal waste such as kitchen garbage and paper waste, and sewage sludge.*" In the absence of a consistent definition for biomass or organic waste, various terms, as noted above, may be found throughout this publication.

Organic waste, if not correctly managed, has the potential to cause significant environmental impacts. As at 2011, South Africa disposed of an estimated 97.3% of all generated organic waste and commercially exploitable biomass to land (including landfill) (DEA, 2012). The decomposition of organic waste in landfill results in the generation of leachate, that permeates down through the waste pile and if not captured through leachate collection systems, has the potential to impact negatively on ground and surface water resources. This decomposition process in landfills also produces landfill gas consisting predominantly of methane (CH_4) and carbon dioxide (CO_2). Methane is a potent greenhouse gas (GHG) with a global warming potential 25 times greater than carbon dioxide. Landfilling of organic waste is therefore a significant contributor to GHG emissions and climate change. The waste sector contributed 3.7% to South Africa's GHG inventory in 2010, up from 2.8% in 2000, predominantly through the disposal of solid waste to landfill. The sector is also the second highest contributor to human-related CH_4 emissions in South Africa, contributing 37.2% to the total CH_4 emissions in 2010, up by 11.3% since 2000 (DEA, 2014). Organic waste can, however, be quite easily diverted away from landfill towards alternative waste treatment options, including composting and anaerobic digestion (AD). However, in recent years, research has shifted to producing higher value end-use products from organic waste, including biochemicals, biopolymers and biomaterials. Many countries are moving towards banning the disposal of organic waste to landfill, or as a minimum, placing severe restrictions on the percentage of biodegradable waste that can be disposed of to landfill, as a means of mitigating the impacts of climate change. South Africa has, to date, only set restrictions on the disposal of garden waste to landfill, with an expected *"25% diversion from the baseline at a particular landfill of separated garden waste"* by 2018, and a 50% diversion by the year 2023 (DEA, 2013).

With respect to biomass and organic waste valorisation, industry has signalled that the South African research community focus their efforts on (i) technology evaluation (specifically technology assessment, demonstration and localisation, technology adaptation and integration, and advisory support to industry on new and emerging technologies); (ii) technology development (specifically scaling-down technologies to mobile, modular units for on-site waste treatment; developing low-technology solutions with high impact; improving technology efficiencies; and upscaling laboratory and pilot scale to full scale); (iii) developing and maintaining highly skilled local capability; and (iv) ensuring needs-driven research in collaboration with industry (DST, 2014).

This book presents just some of the research being undertaken by South African universities and science councils on possible biological, mechanical, chemical and thermal treatment methods, aimed at diverting biomass and organic waste away from landfill, towards value-adding alternatives, in support of the public and private sectors.

Linda Godfrey, Johann Görgens and Henry Roman

REFERENCES

DEA (Department of Environmental Affairs). 2012. National Waste Information Baseline Report. Department of Environmental Affairs, Pretoria.

DEA (Department of Environmental Affairs). 2013. National Environmental Management: Waste Act 59 of 2008. National Norms and Standards for Disposal of Waste to Landfill. Department of Environmental Affairs, Pretoria.

DEA (Department of Environmental Affairs). 2014. GHG National Inventory Report, South Africa, 2000–2010. Department of Environmental Affairs, Pretoria.

DST (Department of Science and Technology). 2013. The Bio-Economy Strategy for South Africa. Department of Science and Technology, Pretoria.

DST (Department of Science and Technology). 2014. A Waste Research, Development and Innovation Roadmap for South Africa (2015–2025). Towards a Secondary Resources Economy. Department of Science and Technology, Pretoria.

DST (Department of Science and Technology). 2014. Industry-Meets-Science Series. Biomass and Organic-Waste: Seeking Alternative Solutions to Disposal. Workshop Proceedings. Department of Science and Technology, Pretoria.

JIE (Japan Institute of Energy). 2008. The Asian Biomass Handbook. A Guide for Biomass Production and Utilization.

UNEP (United Nations Environment Programme). 2009. Converting Waste Agricultural Biomass into a Resource Compendium of Technologies.

Foreword

The Waste Research, Development and Innovation (RDI) Roadmap was approved by the Department of Science and Technology (DST) in 2014, and the Council for Scientific and Industrial Research (CSIR) was appointed as the implementation partner to the DST in 2015.

The Roadmap responds to the research, development, and innovation requirements of the National Waste Management Strategy (NWMS) of the Department of Environmental Affairs (DEA). The NWMS has as its overarching goal the diversion of waste from landfill, taking South Africa from a condition where 90% of solid waste goes to landfill to a condition where only 10% of solid waste goes to landfill. To achieve this requires innovative deployment of technologies as well as non-technological innovations such as improved institutional arrangements.

Organic waste was identified by stakeholders during the development of the national Waste RDI Roadmap as one of five priority waste streams for inclusion in the Roadmap, with the goal statement (beyond 2024) *"Zero organic waste to landfill with maximum value extraction (materials and energy)"*.

The DST hosted an Industry-Meets-Science Workshop on organic waste at the University of KwaZulu-Natal, in Durban from 26 to 27 November 2014. The aim of the workshop was to bring industry and science together to share ideas, which will lead to –

- Identifying the key issues facing industry with respect to biomass- and organic waste
- Jointly scoping new RDI projects
- Increased RDI collaboration between industry and research community, and
- Uptake of RDI outputs by local industries.

The workshop provided a further step towards strengthening the RDI relationship between industry and the research and development (R&D) community. Activities will continue to be supported through the DST Bio-economy Strategy and the Waste RDI Roadmap.

Challenges with regard to technology implementation, as raised by industry during the workshop, include technology demonstration, feasibility and assessment, locally appropriate technologies, scale of technologies and logistics. While identified as challenges, these issues provide opportunities for directed R&D and innovation.

This book is an outcome of the Industry-Meets-Science Workshop on organic waste and presents the original research from South African universities, science councils and research institutions on waste biomass and organic waste valorisation to a broader audience. It is a consolidation of existing R&D in South Africa to disseminate research and experiences on the transformation of waste and biomass to other useful materials with a particular focus on diverting organic waste from landfill.

The DST looks forward to the continued research on waste biomass and organic waste valorisation that will assist in the transition to a greener economy in South Africa.

Mr Imraan Patel
Deputy Director-General: Socio-Economic Partnerships
Department of Science and Technology, South Africa

Acknowledgements

The editors would like to thank the following people and organisations who supported and contributed to this publication:

Authors

Without the ongoing support, enthusiasm and commitment of a group of highly professional scientists and engineers, this publication would not have been possible. We believe that this book reflects the high level of research being undertaken at South Africa's universities and science councils.

External chapter reviewers

All chapters underwent an independent, double-blind review process. The authors and editors are most grateful to the reviewers for the excellent work they did in ensuring that the reviews were conducted to the highest standard, and in providing valuable review comments to the authors.

Prof. M Abdelwahab (Tanta University, Egypt); J Andrew (CSIR, South Africa); A Bissessur (University of KwaZulu-Natal, South Africa); Dr A Bonomi (CNPEM/CTBE, Brazil); Dr A Chimphango (University of Stellenbosch, South Africa); Prof. T de Koker (Durban University of Technology, South Africa); R du Plessis (University of South Africa, South Africa); Dr R Govinden (University of KwaZulu-Natal, South Africa); Dr K Haigh (Stellenbosch University, South Africa); I Kerr (University of KwaZulu-Natal, South Africa); Dr G Kornelius (University of Pretoria, South Africa); Dr D la Grange (University of Limpopo, South Africa); Prof. S Marx (North-West University, South Africa); Dr B Moodley (University of KwaZulu-Natal, South Africa); Dr J Mulopo (University of the Witwatersrand, South Africa); Prof. C Ngila (University of Johannesburg, South Africa); B North (CSIR, South Africa); Prof. S Oelofse (CSIR, South Africa); Prof. B Patel (University of South Africa, South Africa); Dr B Pletschke (Rhodes University, South Africa); Dr P Veluchamy (Ben-Gurion University, Israel); Prof. B Sithole (CSIR, South Africa); Dr J Snyman (Tshwane University of Technology, South Africa); Dr W Stafford (CSIR, South Africa); Prof. A Stark (University of KwaZulu-Natal, South Africa); S Szewczuk (CSIR, South Africa); Dr L Tyhoda (University of Stellenbosch, South Africa); Prof. H von Blottnitz (University of Cape Town, South Africa).

External book reviewers

The book also underwent an independent, double-blind review process. The editors would like to thank the external reviewers.

Funding partners

This publication would not have been possible without the financial support of the Department of Science and Technology, through the Waste Research, Development and Innovation (RDI) Roadmap.

Publishers

The editors would like to thank UNISA Press for believing in the importance of this publication and the need to make this research available to a wide audience.

About the editors

Prof. Linda Godfrey is a Principal Scientist with the Council for Scientific and Industrial Research (CSIR) and Associate Professor at Northwest University, and holds a PhD in Engineering from the University of KwaZulu-Natal. She currently heads the Waste Research Development and Innovation (RDI) Roadmap Implementation Unit on behalf of the Department of Science and Technology (DST), a unit tasked with implementing South Africa's 10-year Waste RDI Roadmap. Her research interests include the role of the waste sector in transitioning South Africa to a green economy; waste innovation; waste economics; the governance, social and behavioural aspects of integrated waste management; the role of waste information as policy and behaviour-change instrument; and the opportunity that waste provides within a circular economy.

Prof. Johann F Görgens is a Professor of Chemical Engineering at the Department of Process Engineering, Stellenbosch University, and holds PhD (Chemical Engineering) and MBA degrees from the same. He leads a research group that performs both development and assessment of process technologies for conversion of biomaterials to valuable chemicals, materials and energy. Of particular interest are wastes and residues from existing activities. Experimental development of processing technologies is undertaken from laboratory to pilot scale, with industrial demonstrations presently being developed. Technology assessments consider the technical performances of processing methods in the context of economic (investment) viability, environmental and social impacts, to provide an integrated view of sustainability of process technologies. Prof. Görgens works with partners in both the public and private sectors on the development and assessment of these technologies to address opportunities for local implementation.

Dr Henry Roman is the Director of Environmental Services and Technologies at the National Department of Science and Technology, South Africa. He holds a PhD in Biotechnology from Rhodes University. His unit is responsible for the oversight of the implementation of the Waste RDI and Water RDI roadmaps, both of which are 10-year strategic plans. He is also responsible for the South African Risk and Vulnerability Atlas (SARVA), a decision-support tool that is managed by the South African Earth Observation Network. He is the current Chair of the International Water Association – South Africa (IWA-SA). He is active in the field of policymaking regarding South Africa's transition to a greener economy and his interests include sustainable development; green economy (circular economy); water innovation; waste innovation; green biotechnology; and complexity theory related to environmental innovation.

Abbreviations and acronyms

ACNC	all-cellulose nanocomposite films
AD	anaerobic digestion
ASTM	American Society for Testing Materials
BMP	biochemical methane potential
BOD	biochemical oxygen demand
CDM	Clean Development Mechanism project
CNC	cellulose nanocrystal/s
CNF	cellulose nanofiber/s
CSIR	Council for Scientific and Industrial Research
DAFF	Department of Agriculture, Forestry and Fisheries
DBSA	Development Bank of South Africa
DEA	Department of Environmental Affairs
DST	Department of Science and Technology
EIA	Environmental Impact Assessment
EMA	Environmental Management Agency
EU	European Union
FT	Fischer-Tropsch
FTIR	Fourier-transform infrared
FTPP	forestry, timber, pulp and paper
GDP	gross domestic product
GHG	greenhouse gas
GN	Government Notice
HCD	human capital development
HPLC	high-performance liquid chromatography
IPCC	Intergovernmental Panel on Climate Change
ISWM	Integrated Solid Waste Management
KZN	KwaZulu-Natal
LAP	laboratory analytical procedure/s
MS	municipal sewage
MSS	municipal sewage sludge
MST	municipal sewage treatment
MSW	municipal solid waste
NCC	nanocrystalline cellulose
NOWCS	National Organic Waste Composting Strategy
NREL	National Renewable Energy Laboratory
N&S	National Norms and Standards for Disposal of Waste to Landfill
NWMS	National Waste Management Strategy
OLR	organic loading rate
PHK	pre-hydrolysis Kraft
RDI	research, development and innovation
R&D	research and development
REIPP	Renewable Energy Independent Power Producer
RSA	Republic of South Africa
SMME	small-, medium- and micro-sized enterprise/s

TS	total solids
TSS	total suspended solids
UNEP	United Nations Environment Programme
VS	volatile solids
WML	waste management licence
WtE	waste-to-energy
WWT	wastewater treatment

QUANTIFYING ORGANIC WASTE
IN SOUTH AFRICA

OVERVIEW OF POTENTIAL SOURCES AND VOLUMES OF WASTE BIOMASS IN SOUTH AFRICA

SHH Oelofse[1,2] and AP Muswema[1]

[1] CSIR Natural Resources and the Environment, PO Box 395, Pretoria, 0001, South Africa

[2] Unit for Environmental Sciences and Management, North-West University, Potchefstroom, 2520, South Africa. Corresponding author e-mail: SOelofse@csir.co.za

ABSTRACT

Global trends in waste management suggest that economic opportunities exist for waste as secondary resources, and waste biomass is no exception. Large quantities of waste biomass are being generated by the agricultural sector, agro-processing and paper and pulp industries in South Africa. The organic fraction of municipal solid waste (MSW) (i.e., garden waste, food waste, paper and wood waste) is also a potential source of waste biomass. There is, however, very little separation at source taking place in South Africa, which could unlock this resource. The agricultural and agro-processing sector utilises waste biomass as animal feed or to generate electricity, mostly for on-site consumption by the local sugar and paper mills. Other opportunities for waste biomass valorisation also exist through the extraction of valuable compounds and the production of liquid fuels through biorefineries. To realise the potential of waste biomass as a secondary resource, it is important to know where the waste is generated, in what quantities it is generated and what is required to unlock these resources. This chapter therefore attempts to shed some light on the available waste biomass resources in South Africa, the relative geographic distribution of the waste and some valorisation opportunities.

Keywords: biomass, quantification, South Africa

INTRODUCTION

Waste biomass is defined by the European Union (EU) as "*the biodegradable fraction of products, waste and residues from biological origin from agriculture (including vegetal and animal substances), forestry and related industries including fisheries and aquaculture, as well as the biodegradable fraction of industrial and municipal waste*" (EU, 2009). This definition includes organic waste (GW20) (i.e. garden waste, food waste and wood waste) and paper (GW50) (i.e. newsprint, brown grades, white grades and mixed grades) as classified in the South African Waste Information Regulations (Regulation 625 of 13 August 2012) (RSA, 2012). For the purposes of this study, the broader EU definition of waste biomass applies.

Landfilling is still considered the most practical and cheapest waste management option in South Africa resulting in the bulk of all waste generated (90%) still being disposed of at landfills (DEA, 2012). However, scarcity of available land close to areas of waste generation, uncontrolled landfill gas (CH_4) and leachate emissions from waste biomass, have caused landfilling to become a less attractive option (Hartmann & Ahring, 2006). Moving waste up the hierarchy towards reuse, recycling and recovery has benefits for the environment and contributes to the principles of a "green economy" in a number of ways, including economic growth and job creation, and reducing social and environmental costs associated with waste management (DST, 2014). Disposal of biodegradable waste to landfill is outlawed in many countries (DEA, 2010) and the phasing out of these practices is now also a priority in South Africa (DEA, 2011).

As at 2011, South Africa generated in the order of 3 million tonnes per annum of biodegradable waste from garden and food waste sources, which is collected as part of the municipal waste stream (DEA, 2012). Of this waste, only a small percentage (35%) was reported to be recycled (DEA, 2012) while the bulk is disposed of at landfills. In addition, 1.7 million tonnes of waste paper and just over 36 million tonnes per annum of commercially exploitable waste biomass including bagasse, wood and sawmill waste and pulp (DEA, 2012) is estimated to be generated in South Africa.

The use of crop residues as resources for bioenergy is attractive since their sustainable use leads to minimal or no land use change impacts (compared to energy crops). According to Batidzirai (2013), maize stover and wheat straw are used for livestock bedding and feed especially in the winter. In South Africa, these residues have the largest residue potential estimated to be in the order of 14.4 million tonnes, of which only 6 million tonnes can be sustainably removed from the fields (Batidzirai, 2013).

Approximately 2.5 million tonnes of food waste is generated in the food processing and packaging stages of the value chain in South Africa (Nahman & De Lange, 2013; Oelofse & Nahman, 2013) and these wastes are also often disposed of at landfills. Fruit and vegetable wastes are rich in nutrients and can be used as animal feed or for the production of value-added products, such as essential oils, polyphenols, anti-carcinogenic compounds, edible oil, pigments, enzymes, bioethanol, biomethane, biodegradable plastic and single cell protein (Choi et al., 2015; FAO, 2013). Fruit waste in particular is not a high value feedstock, due to its relative low protein content, but it is rich in fermentable soluble sugars such as glucose, fructose and sucrose along with structural cellulose and hemicellulose (Choi et al., 2015). These chemical constituents, and the fact that fruit waste is in relative abundant supply, suggest that fruit waste may be a source of waste biomass to consider for ethanol production (Choi et al., 2015).

Given the above potential for South Africa to generate substantial amounts of waste biomass that can be used and valorised, we attempted to identify potential sources and volumes of waste biomass in South Africa. The geographic spread of the waste generators was also considered as it may impact on the accessibility of the waste resources for valorisation opportunities in South Africa.

METHOD

A desktop study of potential sources and quantities of waste biomass from municipal solid waste, food processing, abattoirs (red meat and poultry) and crop residues (wheat and maize) was undertaken. The geographic locations of the food and grain production areas, abattoirs and food processing plants in South Africa were also investigated. Published estimates of percentage waste and by-product generation during different processes for the various commodities were sourced. Mathias *et al.* (2014) report that between 125 kg and 130 kg of wet biomass is generated for every 100 kg of processed grains in the brewery industry. According to Fuentes *et al.* (2004) as quoted by the European Commission (2010), juice production from fruit and vegetables generates between 30% and 50% waste (by weight) while processing and preservation generate between 5% and 30% waste and by-products. The average solid waste generated by slaughter from bovine is 275 kg per tonne of total live weight (i.e. ~ 83 kg per head) (Jayathilakan *et al.*, 2012). In the case of goat and sheep, the average is 17% of animal weight (2.5 kg per head), for pigs 4% (2.3 kg per head) (Jayathilakan *et al.*, 2012) and for poultry 0.41 kg per 1.5 kg bird (Jeon *et al.,* 2013). This information was combined with available statistics on grain production, fruit and vegetable processing and red meat and broiler slaughters in South Africa to calculate estimated volumes of the waste.

RESULTS AND DISCUSSION

Waste biomass from municipal solid waste sources

Waste characterisation studies from various municipalities indicate that between 10% and 54% of municipal solid waste is categorised as organic waste and between 11% and 28% as paper (Table 1.1). The variability of the waste biomass (organic) component in the municipal waste stream could be explained by different climatic conditions in the different municipalities, with municipalities in the drier regions generating less garden waste than areas with higher rainfall. Different socio-economic circumstances between municipalities could also influence waste composition results as well as seasonal variability (CoJ, 2015).

Table 1.1: Domestic waste composition (% by weight) by municipality

Municipality	Organic	Paper	Other	Plastic	Glass	Metals	Reference
City of Cape Town (2010)	39	20	5	18	11	7	DEADP, 2011
Cape Winelands DM (2010)	29	26	13	18	8	6	
Central Karoo DM (2010)	14	28	6	28	13	11	
West Coast DM (2010)	18	19	23	27	6	7	
Overberg DM (2010)	24	22	33	10	5	6	
Eden District (2010)	32	13	2	33	10	10	
Polokwane LM (2007)	40	19	0	18	12	10	Ogola *et al.*, 2011
Sol Plaatjie LM (2010)	10	21	39	18	10	3	SPLM IWMP, 2010
Lejwelepustwa DM (2011)	31	13	27	15	9	5	DEDTEA, 2011
Mangaung MM (March 2010)	38	21	13	14	9	5	MMM, 2014
City of Johannesburg (2014/15)	30	20	29	11	7	3	CoJ, 2015
Msunduzi LM (2010)	36	17	28	10	6	3	UDM, 2010
uMshwathi LM (2010)	37	11	25	9	12	6	
Nelson Mandela Bay MM (2011)	39	15	25	11	8	2	NMBM, 2016
City of Tshwane (2016)	54	12	16	10	7	2	Komen *et al.*, 2016
Average composition across municipalities	**31**	**18**	**19**	**17**	**9**	**6**	

The percentage of food waste in the overall household waste stream in South Africa per income group (annual income) is on average 18.08% for low income (R0-R50k), 10.98% for medium income (R51-R500k) and 9.58% for high income (>R501k) (Nahman *et al.*, 2012) amounting to about overall 15% of the domestic waste generated in South Africa. The contribution of each income group to the total domestic waste stream is estimated to be 58.2% for low income, 30.4% for medium income and 11.4% for high income (Nahman *et al.*, 2012). The total food waste contribution to domestic waste was calculated based on these assumptions and the geographic distribution of the food waste generated in 2011 per province (Table 1.2) was calculated based on the provincial distribution of the adult population per income group in 2010 (BMR, 2010). It is estimated that approximately of 2.4 million tonnes per annum of food waste is generated from domestic sources in South Africa.

The national recovery of organic waste (i.e. garden waste, food waste and wood waste) in 2011 was estimated at 35% or just over 1 million tonnes per annum (DEA, 2012). It is assumed that this 35% recovered organic waste was recycled through composting. It should however be noted that the municipal owned composting facility in Johannesburg has since closed down. There are no accurate records of the current recycling rates of garden waste through composting. Ekelund and Nyström (2007) listed composting projects recovering a total of 225,240 tonnes per annum of green waste from municipal sources producing 113,350 tonnes per annum of compost.

Table 1.2: Food waste generation from domestic sources per province in South
Africa, 2011

Province	Food waste (Tonnes)
Western Cape	246 750
Eastern Cape	346 058
Northern Cape	58 225
Free State	152 130
KwaZulu-Natal	472 164
North West	173 695
Gauteng	510 203
Mpumalanga	178 500
Limpopo	243 371
Total	**2 381 096**

If organic wastes from municipalities are to be considered for valorisation opportunities, it will be advisable to consider strategies for the separate collection (separation at source) of this waste stream from households to reduce the potential of contamination with inorganic materials.

Waste biomass from food processing

Malting barley

The processing of barley into malt is done mainly in Caledon in the Southern Cape, but also in Alrode near Johannesburg. Malting barley only has one major buyer in South Africa, namely South African Breweries Maltings (DAFF, 2014). Based on 2012/13 brewery statistics (DAFF, 2014) and Mathias *et al.* (2014) waste generation figures, it is estimated that between 584,125 and 607,490 tonnes of wet biomass waste was produced. This estimate seems to be in the correct order of magnitude when considering that in 2012 South Africa produced 31.5 million hectolitres of beer (Barthouse Group, 2013) and 14 kg to 20 kg of spent grain is generated per hectolitre of beer produced (Fillaudeau *et al.*, 2006).

Beer brewing generates three waste streams namely spent grain (bagasse), hot trub and residual yeast. All of these residues have a high organic fraction and are high in protein content. The waste streams cannot be reduced and should be considered for use as animal feed, human nutrition or as support for growth of immobilised microorganisms in industrial processes (Mathias *et al.*, 2014). Recent advances in biotechnology suggest that spent brewers' grain can be used as a feedstock for producing several products (Aliyu & Bala, 2011). Potential applications include animal feed and human nutrition, energy and biogas production, protein concentrates and fermentation products including ethanol, lactic acid, gums, antibiotics and enzymes amongst others (Mathias *et al.*, 2014).

Deciduous fruit

Deciduous fruit, including apples, pears, table grapes, peaches, nectarines, apricots and plums, are mainly produced in the Western and Eastern Cape provinces (DAFF, 2014). When applying the average percentage wastage during processing (i.e. 40% for juice and 17.5% for processing and preservation) to the tonnes of fruits processed in South Africa, an estimate of the volume of waste from fruit processing in South Africa can be made (Table 1.3).

It is estimated that approximately 184,290 tonnes of fruit waste was generated from processing of deciduous fruits in 2012/13.

Dried apple pomace is reported to contain 7.7% crude protein and 5.0% ether extract while grape pomace is rich in sugars and phenolics (FAO, 2013). The concentration of total phenolic compounds in the peels, pulp and seeds from apples, peaches and pears, amongst others, is more than twice the amount present in the edible tissue (FAO, 2013). It may therefore be worthwhile to investigate the potential for developing novel value-added products from these waste streams.

Table 1.3: Estimated waste (tonnes) from juice and canning of deciduous fruits in South Africa (2012/13)

Type of fruit	Tonnes processed to juice	Tonnes processed for canning	Waste from juice[†]	Waste from canning[‡]	Total waste from juice and canning (Tonnes)
Apples	259 980.41	26 341.63	103 992.16	4 609.78	108 601.94
Apricots	18 185.50	26 000.00	7 274.20	4 550.00	11 824.20
Pears	47 427.54	83 950.68	18 971.01	14 691.37	33 662.38
Peaches	29 362.84	92 000.00	11 745.14	16 100.00	27 845.14
Grapes	2 356.56	-	942.62	-	942.62
Plums	3 534.84	-	1 413.94	-	1 413.94
Total	360 847.69	228 292.31	144 339.08	39 951.15	184 290.23

† 40% of input by weight
‡ 17.5% of input by weight

Subtropical fruit

The main production areas of subtropical fruit in South Africa are in Limpopo, Mpumalanga and KwaZulu-Natal. Granadillas and guavas are also cultivated in the Western Cape, while pineapples are grown in the Eastern Cape and KwaZulu-Natal (DAFF, 2014). The total production of subtropical fruit in 2012/13 was 675,828 tonnes, of which 131,313 tonnes were taken in for processing (DAFF, 2014). Due to a lack of data on the percentage split between subtropical fruit processed for juice and preservation purposes, the calculations for waste estimates was based on the assumption that 100% of the subtropical fruits taken in for processing is processed for preservation purposes (i.e. 17.5% wasted). The estimated waste from processing of subtropical fruits in South

Africa at an assumed waste generation rate of 17.5% of input, amounted to about 22,980 tonnes in 2012/13 (Table 1.4).

Table 1.4: Estimated waste from processing of subtropical fruit (2012/13)

	Processed (Tonnes)	Waste from processing for preservation[†] (Tonnes)
Avocados	3 922.00	686.35
Bananas	359.00	62.83
Pineapples	68 522.00	11 991.35
Mangoes	28 204.00	4 935.70
Papayas	1 528.00	267.40
Granadillas	45.00	7.88
Litchis	575.00	100.63
Guavas	28 158.00	4 927.65
Total	**131 313.00**	**22 979.78**

†17.5% of input by weight

Citrus fruit

The main production areas for citrus fruit are Limpopo, Eastern Cape, Mpumalanga, Western Cape and KwaZulu-Natal provinces (DAFF, 2014). Citrus production in 2012/13 was 2.3 million tonnes, of which about 23.9% or 560,456 tonnes were taken in for processing (DAFF, 2014). For the purposes of this study, it is assumed that all citrus fruit taken in is processed for juice production. If 40% of the fruit by weight remains as waste, then 224,182 tonnes of waste is generated. Citrus peel waste accounts for almost 50% of the wet fruit mass after processing (Choi *et al.,* 2015). Citrus peel waste is rich in various soluble and insoluble sugars, making it a good candidate as feedstock for bioethanol production, however, it also contains a strong microbial inhibitor namely D-limonene, which first needs to be removed. Choi *et al.,* (2015) states that the production of D-limonene from citrus peel is economically viable, as this by-product has high added value as a flavouring agent and for various applications in the chemical industry.

Vegetables excluding potatoes

Vegetables are produced in most parts of the country with some areas concentrating on specific crops; for example, green beans are grown mainly in Kaapmuiden, Marble Hall and Tzaneen, green peas mainly in George and Vaalharts, onions mainly in Caledon, Pretoria and Brits and asparagus mainly in Krugersdorp and Ficksburg (DAFF, 2014). A total of 2.6 million tonnes of vegetables (excluding potatoes) were produced in 2012/13 of which 7% or 184,450 tonnes were processed (DAFF, 2014). Assuming wastage of 17.5% (processing and preservation), it is estimated that 32,278 tonnes of vegetable waste were generated through processing in 2012/13.

Potatoes

Potatoes are produced, throughout the year, in 16 distinct areas mainly situated in Free State, Western Cape, Limpopo and Mpumalanga (DAFF, 2014). Of the 2.25 million tonnes produced in 2012, approximately 16.5% (about 371,415 tonnes) was taken in for processing. About 91% of the potatoes was processed into potato chips (both fresh and frozen) and the remaining 9% was used for canning, mixed vegetables and other purposes (DAFF, 2014). Potatoes are fat and cholesterol free and high in fibre, Vitamin C and essential minerals like potassium, phosphorus and calcium (DAFF, 2013).

Processing plants peel the potatoes as part of the production of crisps, instant potatoes and similar products. The produced waste is 90 kg per tonne of influent potatoes and is apportioned as 50 kg of potato skins, 30 kg of starch and 10 kg of inert material. The potato peel waste contains sufficient quantities of starch, cellulose, hemicellulose, lignin and fermentable sugars to warrant its use as an ethanol feedstock (Arapoglou *et al.*, 2010). If 371,415 tonnes of potatoes were processed with wastage of 90 kg/tonne, then 33,427 tonnes of waste were generated, of which 56% was potato skins, 33% was starch and 11% was inert material.

Sugar

Sugar cane is mostly produced in KwaZulu-Natal and Mpumalanga with some farms in Eastern Cape (DAFF, 2014). Sugar is manufactured by six milling companies at 14 sugar mills within the cane growing regions. The average production of sugar cane over the past decade (2003/4 to 2012/13) was 18.9 million tonnes per annum and the sugar production for 2012/13 is estimated at 1.95 million tonnes (DAFF, 2014). Sugar cane bagasse is generated at 30% to 40% (by weight) of the input sugar cane for sugar production. In 2012, South Africa produced 6.9 million tonnes of bagasse (UNdata, 2012), which is intensively used in different contexts in South Africa (Carrier *et al.*, 2012). Applications in pulping, activated carbon production, cellulosic ethanol production and energy production through pyrolysis and gasification have been considered in South Africa while in the sugar industry this waste is used for energy generation through combustion (Carrier *et al.*, 2012). According to Smithers (2014, 917) "most sugar factories are designed to burn bagasse inefficiently so as to avoid the need for costly bagasse disposal systems", while others are fitted with back-end refineries (Malelane, Pongola, Umfolozi, Gledhow and Noodsberg). Therefore, improved energy conversion efficiencies at sugar mills in South Africa, could unlock bagasse for other valorisation opportunities without negatively affecting the sugar mills.

Crop residues from wheat and maize production

The amount of biomass from maize stover is estimated to be 16 million tonnes per annum, including aboveground (9.7 million tonnes) and below ground (6.3 million tonnes) biomass (Batidzirai, 2013). Similarly, the wheat straw biomass production is estimated at 1.8 million tonnes (870,000 and 970,000 tonnes of above ground and below ground biomass respectively) (Batidzirai, 2013). It is important to note that not all waste biomass can be removed sustainably, since about 4.2 million tonnes of maize stover will be required for soil erosion control and 9.3 million tonnes for soil organic carbon maintenance (Batidzirai, 2013). The amount of wheat straw required to maintain soil organic carbon and prevent soil erosion amounts to about 870,000 and 100,000 tonnes respectively. In addition, about 70,000 tonnes of straw is utilised as animal bedding (Batidzirai, 2013). The estimated amounts of crop residues from wheat and maize that can be sustainably recovered as sources of biomass for valorisation purposes by province is summarised in Table 1.5.

Table 1.5: Sustainable maize and wheat residue potential by province for South Africa (adopted from Batidzirai, 2013)

Province	Maize stover (Tonnes)	Wheat straw (Tonnes)
Western Cape	10 000	2 000
Northern Cape	1 330 000	385 000
Free State	1 650 000	9 000
Eastern Cape	30 000	3 000
KwaZulu-Natal	31 000	25 000
Mpumalanga	1 580 000	19 000
Limpopo	10 000	99 000
Gauteng	240 000	5 000
North West	10 000	55 000
Total	**5 150 000**	**603 000**

Waste biomass from abattoirs

Red meat and poultry abattoirs account for the bulk of the meat that is legally sold South Africa and therefore the discussion on biomass from abattoirs will be focused on red meat and poultry abattoirs.

Beef is produced throughout South Africa, and the amount produced depends on the available infrastructure, such as feedlots and abattoirs (DAFF, 2012a). Mpumalanga commands the greatest share of beef production in South Africa accounting for 23% of the beef produced in 2011 followed by the Free State (19%), Gauteng (14%), KwaZulu-Natal (12%) and North West (11%) (DAFF, 2012a).

Sheep numbers in South Africa are estimated at 24.5 million distributed in all nine provinces. Approximately 86% of the sheep are produced in Eastern Cape (30%), Northern Cape (25%), Free State (20%) and the Western Cape (11%). The other five provinces share the remaining 14% of the country's sheep numbers (DAFF, 2011a).

The bulk of pork is produced in the Limpopo (24%) and North West (20%) provinces, followed by Gauteng and Western Cape, each producing 11% and KwaZulu-Natal producing 10%, Free State and Mpumalanga 8% each, Eastern Cape 6% and Northern Cape 2% (DAFF, 2011b).

Broiler meat is produced throughout South Africa, with North West, Western Cape, Mpumalanga and KwaZulu-Natal provinces being the largest producers accounting for approximately 81% of total production (DAFF, 2012b). The average number of animals slaughtered per week in 2015 was calculated based on statistics from the Red Meat Abattoir Association (RMAA) for bovine, sheep and pigs (RMAA, 2015). The estimated amount of solid waste generated at red meat abattoirs in 2015 was 40,465 tonnes (Table 1.6).

Table 1.6: Estimated quantity of solid waste generated by slaughter of bovine, sheep and pigs

Meat source	Average slaughters per week [a]	Waste Quantity (kg/head) [b]	Waste generated (Tonne/week)	Waste generated (Tonne/ annum)
Bovine	8 822.00	83.00	732.23	38 075.75
Sheep	11 817.00	2.50	29.54	1 536.21
Pigs	7 132.00	2.30	16.40	852.99
Total	-	-	778.17	40 464.95

(a) RMAA, 2015
(b) Jayathilakan et al. 2012

In 2013, on average 18,169 million broilers were slaughtered per week (947,421 million per annum) in South Africa (South African Poultry Association, 2013) generating about 7.6 million tonnes of solid waste (including giblets) per week (Table 1.7).

Table 1.7: Estimated quantity of solid waste generated by poultry slaughter

By-product	Quantity per head of chicken (kg) [a]	Waste per week (Tonnes)	Waste per annum (Tonnes)
Giblets	0.13	2 400 000	125 000 000
Heads	0.04	800 000	42 000 000
Blood	0.03	620 000	32 000 000
Feathers	0.09	1 640 000	85 000 000
Sludge cake	0.12	2 180 000	8 470 000 000

(a) Jeon et al., 2013

Waste biomass from abattoirs is used in various sectors including the medical and pharmaceutical industry (e.g. blood, collagen and hormones), food industry (e.g. gelatin, collagen), animal feed (e.g. meat and bone meal), pet food and fertilisers (Jayathilakan *et al.*, 2012).

Table 1.8: Summary of waste biomass sources identified in South Africa

Sector	Waste material available	Location	Estimated biomass waste available in South Africa
Municipal solid waste	Garden waste	All municipalities	No estimate could be done due to lack of data
	Food waste from domestic sources	All municipalities	2.38 million tonnes (2011)
	Paper	All municipalities	On average 19% of domestic waste
Malting barley	Spent grain	Caledon, Alrode	584,125 - 607,490 tonnes of wet bagasse (2012/13)
Deciduous fruit	40% waste for juice and 17.5% waste for processing and preservation	Western Cape Eastern Cape	184,290 tonnes of fruit waste in 2012/13 (144, 339 tonnes from juice and 39 951 tonnes from canning)
Subtropical fruit		Limpopo Mpumalanga KwaZulu-Natal Western Cape	22,980 tonnes (2012/13)
Citrus fruit	Peels (50% of wet juiced fruit)	Limpopo, Eastern Cape, Mpumalanga, Western Cape, KwaZulu-Natal	224,182 tonnes (2012/13)
Vegetables (excluding potatoes)		Kaapmuiden, Marble Hall and Tzaneen, green peas mainly in George and Vaalharts, onions mainly in Caledon, Pretoria and Brits and asparagus mainly in Krugersdorp and Ficksburg	32,278 tonnes (2012/13)
Potatoes	Skins (56% of processed) Starch (33% of processed) Inert material (11% of processed)	Free State, Western Cape, Limpopo, Mpumalanga	33,427 tonnes (2012)
Sugar	Bagasse	KwaZulu-Natal, Mpumalanga, Eastern Cape	6.90 million tonnes (2012)
Crop residues from	Maize stover	Northern Cape, Free State, Mpumalanga	5.15 million tonnes/annum
	Wheat straw	Northern Cape, Limpopo, North West	0.60 million tonnes/annum
Abattoirs	Red meat	Mpumalanga, Free State, Gauteng	40,464.95 million tonnes/annum
	Poultry	North West, Western Cape, Mpumalanga, KwaZulu-Natal	272.1 million tonnes/annum

CONCLUSIONS

Based on this desktop study of publicly available sources (Table 1.8), between 10% and 54% of municipal solid waste from households is biomass, which is believed to be food waste and garden waste of variable composition, and between 11% and 28% is paper of different grades. The observed variability could be attributed to varying socio-economic and climatic conditions (for instance rainfall); in the different municipalities, as well as seasonality, since most data from waste characterisation studies reviewed refer to information provided for one season only. Food waste was estimated to be in the order of 2.38 million tonnes/annum (or ~15% of the total domestic waste generated in 2011). The magnitude of waste biomass from domestic sources suggest that municipal solid waste presents an opportunity for biomass valorisation, however, this source will be challenging to capitalise given the poor level of separation at source by households and relative low participation rates in such initiatives currently experienced in South African municipalities.

Waste biomass from food processing presents an opportunity to utilise point sources of waste biomass for valorisation. The most obvious of these is the sugar industry, which produced an estimated 6.9 million tonnes of bagasse waste in 2012. Currently, the majority of this waste material is inefficiently utilised for energy generation by the sugar industry but other applications are also possible, especially if energy conversion efficiencies at sugar mills could be improved.

In comparison, food waste from food processing facilities may present a more immediate opportunity that can be leveraged through direct interaction with the industry. The waste streams generated are likely to be homogeneous and easily accessible. Fruit and vegetable waste therefore has potential as livestock feed and as substrates for the generation of other value-added products (FAO, 2013). Valorisation of waste biomass should therefore consider maximum value addition to the waste to maximise the value that can be extracted from the waste as secondary resources.

Waste from abattoirs is also a significant source. It should be noted that some of this waste is already used, for instance, in the medical and pharmaceutical industry, food industry and as animal and pet food. However, large quantities of abattoir waste are still disposed of at landfills, causing environmental impacts. Therefore, any valorisation opportunity for this waste stream will be welcomed.

The bulk (by mass and possibly volume) of solid waste disposed of at municipal landfills in South Africa is waste biomass. Diverting these wastes from landfill through valorisation will save landfill airspace, significantly reduce greenhouse gas (GHG) emissions from landfills and contribute to the transition to a green economy in South Africa by realising the resource potential of waste biomass.

REFERENCES

Aliyu, S and Bala, M. 2011. Brewer's spent grain: A review of its potentials and applications. *African Journal of Biotechnology,* 10(3), 324–331.

Arapoglou, D, Varzakas, TH, Vlysside, A and Israilides, C. 2010. Ethanol production from potato peel waste. *Waste Management* 30, 1898–1902.

Barthause Group. 2013. Beer production market leaders and their challenges in the Top 40 Countries in 2012.

Batidzirai, B. 2015. Design of sustainable biomass value chains: Optimising the supply logistics and use of biomass over time. Ph.D. dissertation. Department of Innovation, Environmental and Energy Sciences, Copernicus Institute of Sustainable Development, Utrecht University, the Netherlands.

BMR (Bureau of Market Research). 2010. Personal income estimates in South Africa, 2010. BMR Report 396.

Carrier, M, Hardie, AG, Uras, Ü, Görgens, J and Knoetze, J. 2012. Production of char from vacuum pyrolysis pf South African sugar cane bagasse and its characterisation as activated carbon and biochar. *Journal of Analytical and Applied Pyrolysis,* 96, 24–32.

Choi, IS, Lee, YG, Khanal, SK, Park, BJ and Bae, H-J. 2015. A low-energy, cost effective approach to fruit and citrus peel waste processing for bioethanol production. *Applied Energy,* 140, 65–74.

CoJ (City of Johannesburg). 2015. Feasibility study for alternative waste treatment technology: Part 3; Waste Characterisation study for CoJ including Sept 2014, Nov 2014 and June 2015 sampling.

DAFF (Department of Agriculture, Forestry and Fisheries). 2011a. A Profile of the South African mutton market value chain. Department of Agriculture, Forestry and Fisheries, Pretoria.

DAFF (Department of Agriculture, Forestry and Fisheries). 2011b. A Profile of the South African pork market value chain. Department of Agriculture, Forestry and Fisheries, Pretoria.

DAFF (Department of Agriculture, Forestry and Fisheries). 2012a. A Profile of the South African Beef Market Value Chain, 2012. Department of Agriculture, Forestry and Fisheries, Pretoria.

DAFF (Department of Agriculture, Forestry and Fisheries). 2012b. A Profile of the South African Broiler Market Value Chain, 2012. Department of Agriculture, Forestry and Fisheries, Pretoria.

DAFF (Department of Agriculture, Forestry and Fisheries). 2013. A Profile of the South African Potato Market Value Chain, 2012. Department of Agriculture, Forestry and Fisheries, Pretoria. Available at: http://www.nda.agric.za/docs/AMCP/POTATO2012.pdf. [Accessed 28 October 2015].

DAFF (Department of Agriculture, Forestry and Fisheries). 2014. Trends in the Agricultural Sector, 2013. Department of Agriculture, Forestry and Fisheries, Pretoria, 69.

DEA (Department of Environmental Affairs). 2010. National waste management strategy - First draft for public comments, Department of Environmental Affairs, Pretoria.

DEA (Department of Environmental Affairs). 2011. National Waste Management Strategy, November 2011. Department of Environmental Affairs, Pretoria.

DEA (Department of Environmental Affairs). 2012. National Waste Information Baseline Report. Department of Environmental Affairs, Pretoria.

DEADP (Department of Environmental Affairs and Development Planning, Western Cape). 2011. Status Quo Report: Integrated Waste Management Plan for the Western Cape Province. February 2011.

DEDTEA (Free State Department of Economic Development, Tourism and Environmental Affairs). 2011. Lejweleputswa District Municipality Integrated Waste Management Plan. Final IWMP, August 2011.

DST (Department of Science and Technology). 2014. A National Waste R&D and Innovation Roadmap for South Africa: Phase 2 Waste RDI Roadmap. The economic benefits of moving up the waste management hierarchy in South Africa: The value of resources lost through landfilling. Department of Science and Technology, Pretoria.

Ekelund, L and Nyström, K. 2007. Composting of municipal waste in South Africa – Sustainability aspects. Uppsala University. Available at: www.utn.uu.se/sts/cms/filarea/0602_kristinanystromlottenekelund.pdf. [Accessed 19 April 2013].

European Commission. 2010. Preparatory study on food waste across EU 27. Technical Report – 2010-054.

EU (European Union). 2009. Directive 2009/28/EC of the European Parliament and of the Council of 23 April 2009 on the promotion of the use of energy from renewable sources and amending and subsequently repealing Directives 2001/77/EC and 2003/30/EC.

FAO (Food and Agricultural Organisation of the United Nations). 2013. Utilisation of fruit and vegetable wastes as livestock feed and as substrates for generation of other value added products. RAP publication 2013/04. ISBN 978-92-5-107631-6.

Fillaudeau, L, Blanpain-Avet, P and Daufin, G. 2006. Water, waste water and waste management in brewing industries. *Journal of Cleaner Production,* 14, 463–471.

Hartmann, H and Ahring, BK. 2006. Strategies for the anaerobic digestion of the organic fraction of municipal solid waste: an overview. *Water Science and Technology*, 53(8), 7–22.

Jayathilakan, K, Sultana, K, Radhakrishna, K and Bawa, AS. 2012. Utilisation of byproducts and waste materials from meat, poultry and fish processing industries: a review. *Journal of Food Science Technology,* 49(3), 278–293.

Jeon, Y-W, Kang, J-W, Kim H, Yoon, Y-M and Lee, D-H. 2013. Unit mass estimation and characterisation of litter generated in the broiler house and slaughter house. *International Biodeterioration and Biodegradation,* 85, 592–597.

Komen, K, Mtembu, N, Van Niekerk, MA and Perry, EJ. 2016. The role of socio-economic factors, seasonality and geographic differences on household waste generation and composition in the City of Tshwane. In *Proceedings of the 23rd WasteCon Conference* 17–21 October 2016, Johannesburg, South Africa.

Mathias, TRDS, De Mello, PPM and Servulo, EFC. 2014. Solid wastes in brewing processes: a review. *Journal of Brewing and Distilling* 5(1), 1–9. July 2014.

MMM (Mangaung Metropolitan Municipality). 2014. Review of 2011 Integrated Waste Management Plan. Draft Situational Analysis, July 2014.

Nahman, A and De Lange, W. 2013. Cost of food waste along the value chain: evidence from South Africa. *Waste Management,* 33, 2493–2500.

Nahman, A, De Lange, W, Oelofse, S and Godfrey, L. 2012. The costs of household food waste in South Africa. Waste Management, 32(11), 2147–2153.

NMBM (Nelson Mandela Bay Municipality). 2016. Nelson Mandela Bay Municipality Integrated Waste Management Plan 2016–2020.

Oelofse, SHH and Nahman, A. 2013. Estimating the magnitude of food waste generated in South Africa. Waste *Management and Research,* 31(1), 80–86.

Ogola, JS, Chimuka, L and Tshivhase, S. 2011. Management of municipal solid wastes: A case study in Limpopo Province, South Africa, in: Kumar, S. (ed.). Integrated Waste Management, Volume 1, InTech, Rijeka, Croatia.

RMAA (Red Meat Abattoir Association). 2015. National South African Price Information statistics. Available at: http://www.rpo.co.za/InformationCentre/IndustryInformation/SlaughteringStatistics. aspx. [Accessed 27 June 2016].

RSA (Republic of South Africa). 2012. National Waste Information Regulations, 2012 (Regulation 625 of 13 August 2012). *Government Gazette 35583 of 13 August 2012.*

Smithers J. 2014. Review of sugar cane trash recovery systems for energy cogeneration in South Africa. *Renewable and Sustainable Energy Reviews,* 32, 915–925.

South African Poultry Association. 2013. Poultry meat industry stats: 2013. Report of the broilers organisation committee. Available at: http://sapoultry.co.za/pdf-statistics/Poultry-meat-industry-stats-summary.pdf. [Accessed 27 June 2016].

SPLM (Sol Plaatjie Local Municipality). 2010. Sol Plaatjie Local Municipality Integrated Waste Management Plan. Final.

UDM (uMgungundlovu District Municipality). 2010. Advanced Integrated Waste Management System for uMgungundlovu District Municipality – Waste Characterisation Study.

UNData. 2012. Energy Statistics Database, United Nations. Available at: www.data.un.org [Accessed 28 October 2015].

2

IDENTIFICATION AND CHARACTERISATION OF TYPICAL SOLID BIOWASTE RESIDUES IN SOUTH AFRICA: POTENTIAL FEEDSTOCKS FOR WASTE-TO-ENERGY TECHNOLOGIES

N Tawona,[1,2] **BB Sithole**[1,2] **and J Parkin**[3]

[1] Discipline of Chemical Engineering, University of KwaZulu-Natal, 358 King George Ave, Durban, 4041, South Africa

[2] Biorefinery Industry Development Facility - CSIR/UKZN, 35 King George Ave, Durban, 4041, South Africa

[3] Durban Cleansing and Solid Waste (DSW), eThekwini Municipality, 7 Electron Road, Durban, 4001, South Africa

Corresponding author e-mail: bsithole@csir.co.za

ABSTRACT

The suitability of different technology solutions for the beneficiation of biowaste is dependent on the physical and chemical properties of the specific waste stream. This research was undertaken as part of the EU FP7 funded Biowaste4SP project, which entailed beneficiation of biowaste via generation of biogas and production of compost – thus benefitting communities in developing countries in terms of avoidance of landfilling and generation of high value materials from biowaste. A study was conducted to identify and characterise the various solid biowaste streams available in South Africa with the ultimate aim of converting the waste into valuable products via waste-to-energy technologies to avoid their disposal by landfilling. Samples were collected from within the KwaZulu-Natal, Gauteng and Limpopo provinces. Representative samples for the identified streams were collected for analysis. The analytical tests performed for the characterisation of the solid biowaste residues included proximate and ultimate analysis, moisture content determination, ash content analysis, and strong acid hydrolysis for determination of total carbohydrates and acid insoluble material. Twelve biowaste types were identified. Sawdust had the highest volatile solids (VS) content (93%) whereas soybean waste and water hyacinth had the lowest VS content (63%). Elemental analysis results revealed that sugar cane bagasse, vegetable waste and banana (whole fruit) were highly carbonaceous (C > 40%) and furthermore, that sugar cane bagasse had the highest carbon : nitrogen (C : N) ratio (246.6:1). The results indicated that banana biowaste is suitable for conversion into biogas.

Keywords: bioresidues, biowaste, volatile solids, waste-to-energy

INTRODUCTION

In an effort to support the currently depleting fossil fuel resources and to generate alternative fuels, many different studies relating to the subject of biomass potential and biomass utilisation have been conducted (Raclavska *et al.*, 2011). Biomass fuels come from a wide range of raw material sources. Presently, the biomass classes identified include virgin wood, energy crops, industrial wastes, food wastes and agricultural residues (Barletta *et al.*, 2015). Raw biomass is not considered an ideal feedstock for biorefinery processing to energy technologies due to its low energy density resulting from both its physical form and moisture content (Zhijia *et al.*, 2016). This makes raw biomass inefficient for storage and transport. Furthermore, it is usually unsuitable for use without extensive pre-treatment, thus altering overall process economics. Liquid biofuels provide one of the few options for fossil fuel replacement in the short to medium term. Liquid biofuels have the potential to offer both greenhouse gas savings and energy security; however, in mid-2008, rising food prices and reported poor energy balances – particularly from first generation biofuel crops – led to their use being questioned at both local and global levels (Taylor, 2008). Furthermore, land displacement is a cause for concern. This focuses attention on food waste, industrial waste, as well as agricultural residues. From a South African perspective, utilisation of waste to produce sustainable products is still in the infancy stage, even though legislation has been amended to include policies that support achievement of zero-waste and encourage waste-to-energy (WtE) initiatives (DEA, 2012). There is a need to identify the waste streams available in South Africa and investigate their potential as far as sustainable product development is concerned. There is still insufficient data on the available biowaste feedstocks in most developing nations. Research shows that either the waste generation data is not catalogued at all, or it is incorrectly and unreliably recorded due to a lack of a properly functioning waste management systems. It is thus paramount to carry out several data collection studies to indicate the waste generation trends. In developing nations, waste collection systems are not as effective as in developed countries.

The most common and cost effective route for waste disposal in South Africa is landfilling. This practice is associated with a number of problems, the most significant of which are environmentally related, for example, greenhouse gas (GHG) emissions. The lack of sufficient scientifically collected information and research conducted on the topic of biowaste potential affects development of legislation that supports waste utilisation. The organic fraction of solid waste can be processed, for example, in biogas plants for renewable energy generation, or in composting facilities for biofertiliser production. However, for these processing facilities to operate in a sustainable manner there is a need to quantify and characterise the available waste residues, and to evaluate their potential to serve as feedstocks for sustainable product development (Lestander, 2010). Energy value can be estimated by analysis of the physico-chemical properties of the waste residues. Ultimate analysis and chemical composition data can be used to estimate the theoretical biogas and methane (CH_4) production potential of a particular waste substrate. The same data can be used to find the perfect combinations for co-composting of dry and green waste residues to produce a cheap form of biofertiliser, among other applications. Most African nations have the potential to have biobased economies by adoption of the biorefinery concept (Pradhan & Mbohwa, 2014). The approach adopted in this study was to identify and characterise the potential biowaste feedstocks that could be converted into value-added products. Examples of value-added products include biogas, biofertiliser, ethanol, lactic acid, amino acid and protein. The identification of the feedstocks was done based on the classification of the waste material as "nutrient-rich" and/or "sugar-rich" (Gustavsson *et al.*, 2014). This classification is very important for product end use. Sugar-rich biowaste can be processed into high value products, such as pharmaceuticals, food and feed ingredients, as well as energy and bulk chemicals whereas nutrient-rich residues can be converted more suitably into fertiliser and soil conditioners. In this study, results obtained from biowaste characterisation were presented.

MATERIALS AND METHODS

Identification of biowaste residues

The first step was the identification of the biowaste residues. Samples were collected from identified sampling areas. Samples that were collected within the eThekwini Municipality were identified by municipal engineers who have some data and knowledge of waste distribution within the municipality. Cattle manure samples were collected from beef cattle farms around the Pinetown area in the KwaZulu-Natal (KZN) province. Fruit and vegetable residues were sourced from fresh produce farms in Verulam, KZN. Garden refuse samples were obtained from the Glanville Garden refuse site in Durban, KZN. This refuse site services a number of suburbs; hence, a representative sample was collected. Sawdust and wood bark were collected from a paper mill in KZN that processes eucalyptus hardwood. Soybean waste residues were collected from soy farms around Pinetown. Sugar cane bagasse was obtained from a sugar mill in KZN. Water hyacinth, an undesirable species that grows uncontrollably and chokes the rivers at some locations (UNEP, 2012), was obtained from the Umlazi River. Banana samples and maize bran residues were obtained from the Limpopo province. Wheat bran residues were collected from Pretoria West in Gauteng province.

Sampling methods and sample preparations

Sampling of biowaste residues was performed following a protocol prepared during the first six months of the Biowaste4SP project (Sundqvist *et al.*, 2013). This protocol was based on standard scientific sampling methods. Primary samples were obtained for each of the identified biomass feedstocks. The primary samples were collected in such a manner that different parts of the volume to be sampled were randomly covered. These sub-samples were mixed together. Coning and quartering were used to reduce the size of the mixed samples. The samples were dried in an oven at 60°C until constant weight was achieved. Milling was then performed to reduce sample particle size to < 1 mm. Equipment combination of a TRF 400 hammer mill and a laboratory blender was used for size reduction. Banana waste was analysed in two ways, banana "fruit-only" and "whole banana" (comprising of fruit and peel). The prepared samples were packed in plastic sample bags and stored at 4°C for analysis.

Methods for characterisation of the sampled biowaste residues

The samples were characterised by proximate and ultimate analyses. In general, proximate analysis classifies the biowaste feedstocks in terms of their moisture, ash, volatile matter and fixed carbon content, however, in this work, fixed carbon was not determined, as this was not within the scope of the Biowaste4SP project. Ultimate analysis was performed and elemental carbon (C), hydrogen (H), nitrogen (N), oxygen (O), and sulphur (S) compositions were measured.

Laboratory analytical procedures based on protocols developed by the National Renewable Energy Laboratory (NREL) were followed for the characterisation of biomass in the Biowaste4SP project with respect to proximate analysis. Furthermore, standard methods were used for the determination of sodium (Na), phosphorus (P), and potassium (K) (Eaton *et al.*, 1998). Methods used for dry, ash and carbohydrate determination are described in detail in the following section.

Determination of dry matter content

The prepared samples were weighed before and after drying in an oven at 105°C overnight. The difference in weight indicated the dry matter content, which was then expressed as a percentage of the initial weight of sample and denoted as DM (%).

Determination of ash content

The prepared samples were weighed before and after "ashing" in a muffle furnace at 550°C for two hours. The difference in weight indicated the ash content, which was then expressed as a percentage of the initial weight of sample, denoted as ash (%). The difference between DM (%) and ash (%) indicated volatile solids (VS) content, expressed as VS (%).

Determination of carbohydrate content

Standard methods developed for high-performance liquid chromatography (HPLC) were used for determination of carbohydrate content of the biowaste residues identified. Strong acid hydrolysis was used to separate the monomer sugars that were then quantified using HPLC analysis. Solvent extraction was first performed on the prepared samples to remove lipids. The solvent-free samples were then dissolved in 72% (w/w) sulphuric acid (H_2SO_4) for one hour in glass test tubes immersed in a water bath maintained at 30°C. The dissolved samples were then hydrolysed by autoclave at 121°C for one hour. Acid insoluble material was measured as the dry residue obtained after filtration and overnight drying of the hydrolysed samples, respectively. Finally, determination of the carbohydrates was performed using an HPLC system equipped with a 4×250 mm Dionex™ CarboPac™ PA1 column maintained at 30°C throughout the run, using a pulsed amperometric detector (PAD).

RESULTS AND DISCUSSIONS

Classification of typical biowaste residues

In South Africa, waste generation data is not readily available and thus classification is still limited to general waste, hazardous waste, and building and demolition waste according to the National Waste Information Baseline Report (DEA, 2012). The identified typical biowaste residues in South Africa in this work were classified as indicated in Table 2.1.

Table 2.1: Waste classification of typical biowaste residues identified in South Africa

Waste category	Biowaste residue	Estimated annual production (Million tonnes)
Food waste	Banana (fruit only)	10.2*
	Banana (whole fruit)	
	Fruit & vegetable waste	
Agricultural waste	Wheat bran	-
	Maize bran	
	Cattle manure	
	Garden waste	
	Soybean waste	
Industrial waste	Sawdust	0.8*
	Sugar cane bagasse	3.2*
	Wood bark	-

Oelofse & Muswema (2013) – South Africa country report

The biomass residues in Table 2.1 are typical biowaste materials available in South Africa. From Table 2.1, it can be seen that the estimated amounts of the individual waste streams were not completely recorded. This is because in South Africa, there is still a lack of properly catalogued data and research as far as solid waste generated across all spectrums is concerned. The research in this report was not focused on quantifying the amounts of waste. Rather, it represents the first stage of the identification, estimation and cataloguing of the potential biowaste feedstocks that are available in South Africa. Furthermore, the results presented in this work emanated from the Biowaste4SP project, whose mandate was to identify and characterise certain feedstocks that were common in the participating African countries, namely South Africa, Egypt, Ghana, Morocco and Kenya. The banana peel was therefore not analysed separately; instead, the fruit and peel were combined for comparison of results from the other African countries where banana waste is more prevalent than in South Africa. Samples were selected from three provinces in South Africa, namely KZN, Gauteng and Limpopo, with most of the samples collected from KZN, mainly due to the costs associated with sampling in all provinces. However, the samples obtained were considered "typical" biowaste samples found in South Africa. The banana and maize bran samples were sourced from Limpopo, whereas wheat bran was sourced from Gauteng for ease of transportation and analysis at the CSIR laboratories in Pretoria. The study focused on household biowaste and not industrial waste streams since the later exhibit a great deal of homogeneity, as they are produced from standardised industrial processes, for example, from pulp and paper mills, or from sugar cane processing mills. The organic/inorganic inconsistencies due to geographical location were thus assumed to be negligible.

Proximate and ultimate analysis

Results from proximate and ultimate analyses are presented in this section. Table 2.2 shows dry matter and ash content results. Table 2.3 shows carbohydrate content as well as acid insoluble material. For lignocellulosic residues, the acid insoluble material was quantified as Klason lignin, defined as the lignin component that is insoluble in 72% sulphuric acid (Theander *et al.*, 1995).

Table 2.2: Dry matter and ash content of the identified biowaste in South Africa

Biowaste residue	Source location	DM (%)	Ash (%)	VS (%)
Sugar cane bagasse	KZN	95.9	5.6	90.3
Wood bark	KZN	94.3	6.2	88.1
Water hyacinth	KZN	93.9	30.7	63.2
Soybean waste	KZN	93.9	30.7	63.2
Sawdust	KZN	93.5	0.6	92.9
Cattle manure	KZN	93.2	31.2	62.0
Maize bran	Limpopo	91.2	3.9	87.3
Garden waste	KZN	90.3	10.8	79.5
Wheat bran	Pretoria West	89.1	5.7	83.4
Banana (fruit only)	Limpopo	88.2	6.9	81.3
Banana (whole fruit)	Limpopo	87.8	16.5	71.3
Fruit & vegetable waste	KZN	86.6	9.5	77.1

DM = Dry matter; VS = Volatile solids; KZN = KwaZulu-Natal

It should be noted that the results in Table 2.2 represent characteristics of the dried and prepared samples, and not raw residues.

All of the residues characterised contained dry matter above 85%. This result, however, does not indicate the actual dry material of the raw samples, as during preparation the samples were dried at 60°C before analysis. The ash content has a direct relation to mineral content (Gustavsson *et al.*, 2014). Soybean waste, banana (whole fruit) and water hyacinth had higher ash contents compared to the other biowaste residues. This directly relates to the minerals' abundance in these waste types. On the other hand, sawdust showed the lowest ash content (0.62%).

Volatile solids are the organic portion that can be converted into products after treatment (combustion, anaerobic digestion, composting) (Eliyan, 2007). Results in Table 2.2 indicate that all the biowaste materials have a volatile solids content greater than 50% with sawdust having the highest VS (92.9%) and water hyacinth and soybean waste tying with the lowest VS content of 63.2%.

Results obtained from carbohydrate characterisation and ash content can be used to classify the waste residues as either "nutrient-rich" or "sugar-rich" waste biomass. As mentioned earlier, high ash content represents high nutrient content. This is important when deciding the best way for beneficiation of the waste materials. Nutrient-rich feedstocks are far more suitable for use as composting materials or directly as soil conditioners than nutrient-deficient feedstocks (Gustavsson et al., 2014).

Sugar-rich feedstocks were designated based on their glucan content. Glucan cntent largely represents the sum of cellulose and starch (Gerardi, 2003). Of the 12 biowaste residues identified in this work, six fell in the group with 10% to 30% glucan, five biowastes had a glucan content between 30% and 50%, and only one (sugar cane bagasse) had a glucan content higher than 50%. Sugar cane bagasse is expected to have a high glucan (sugar) content since it is derived from sugar cane (Table 2.3).

It is important to note that the amount of Klason lignin or acid insoluble material will have a bearing on treatment and conversion technologies of biowastes. Sawdust, cattle manure, wood bark, water hyacinth and garden waste had higher amounts (25%—50%) of acid insoluble material than the other biowastes. This will affect beneficiation of the biowastes, for example, if the waste is to be processed in a fermentation system, pretreatment of the waste becomes essential for removal of lignin, thus affecting economic viability of the beneficiation process. It is worth noting that sugars can further be split into "simple sugars, "starch-rich sugars," or "lignocellulosic sugars" (Gustavsson et al., 2014). Starch-rich residues and simple sugars are associated with first generation technologies whereas lignocellulosic materials require second-generation technologies to access the sugars that would then be converted into valuable products (Luque et al., 2008).

In addition to carbohydrate characterisation, nutrient analysis (protein, starch and total fat) was carried out for all the biowaste residues (Table 2.4) (Eaton et al., 1998). Nutrient analysis of biomass waste is crucial in developing models to predict product end use. Since fats, proteins and carbohydrates are the major building blocks of organic materials, these are the same components that will be converted into value-added products, and the relative proportions of each building block in a particular biowaste type can greatly influence the valorisation of that particular waste.

Table 2.3: Carbohydrates and Klason lignin/acid insoluble material in identified biowastes

Waste material	Source	*KL (%)	Arabinan (%)	Galactose (%)	Rhamnose (%)	Glucan (%)	Xylan (%)	Mannose (%)
Wood bark	KZN	48.3	2.88	3.40	0.00	22.6	3.15	1.94
Garden waste	KZN	37.1	2.98	2.96	0.49	18.9	5.60	0.20
Sawdust	KZN	34.4	0.18	0.51	0.00	47.5	7.40	0.73
Water hyacinth	KZN	26.0	4.82	3.29	0.54	14.3	3.13	0.02
Cattle manure	KZN	25.0	1.52	1.10	0.80	14.6	8.59	0.23
Sugar cane bagasse	KZN	18.1	0.14	0.02	0.00	65.8	5.89	0.00
Fruit & vegetable waste	KZN	13.4	1.70	2.62	0.88	22.5	1.44	0.33
Banana (whole fruit)	Limpopo	12.8	1.12	0.94	0.00	35.4	0.64	2.71
Soybean waste	KZN	11.3	1.48	5.52	0.00	29.4	8.81	1.10
Banana (fruit only)	Limpopo	10.8	0.47	0.65	0.00	38.3	0.60	3.39
Wheat bran	Pretoria West	9.3	7.19	2.07	0.00	34.8	10.6	0.00
Maize bran	Limpopo	1.73	8.23	3.48	0.00	46.5	10.8	0.00

*KL – Klason lignin (Also indicates acid insoluble material where KL is not applicable, e.g., banana fruit)

Table 2.4: Nutrient analysis results for identified biowastes

	Protein (N×6.25) (g/100g)	Starch (g/100g)	Total Fat (g/100g)
Banana (fruit only)	13.1	12.9	6.2
Wheat bran	16.4	14	3.7
Fruit & vegetable waste	16.3	1.7	3.4
Banana (whole fruit)	8.1	0.8	1.4
Garden waste	7.7	0.2	0.6
Soya waste	7.3	0.4	0.6
Wood bark	2.2	0.3	0.6
Cattle manure	11.5	<0.1	0.5
Water hyacinth	14.1	0.2	0.4
Maize bran	5.2	3.3	0.4
Sugar cane bagasse	1.6	0	0.2
Sawdust	0.4	0	0.1

One main objective of the Biowaste4SP project was valorisation of biowaste into biogas. Table 2.5 illustrates the relative contributions of carbohydrates, fats, and proteins to biogas and CH_4 production according to Baserga (1998).

Table 2.5: Gas yield and CH_4 quality of different organic components (Baserga, 1998)

Nutrient	Gas yield (L/kg organic matter)	CH_4 (%)
Carbohydrates	770	50
Fats	1250	68
Proteins	700	71

From the table it is evident that the highest biogas and highest CH_4 quality are expected from fats. Biogas yields from proteins and carbohydrates are comparable, however; CH_4 quality was higher from proteins. The results in Table 2.4 indicated that banana (fruit only) had the highest fat content (6.2g/100g) whereas wheat bran, fruit and vegetable waste, water hyacinth, banana (fruit only) and cattle manure had considerably high protein content (> 10g/100g). However, their fat and carbohydrate contents were lower, with the exception of banana fruit, which showed higher nutritional value than all the other waste residues identified. The nutrient analysis results indicated that banana (fruit only) could be an excellent candidate for biogas production. However, this choice could pose problems since bananas are food material and using them in WtE plants could negatively affect their use in the food industry. Wheat bran on the other hand had a relatively high nutrient content and could be a perfect candidate for biogas production.

Since biowastes fruit and vegetable waste, banana (whole fruit), sugar cane bagasse and cattle manure were common in the five participating African countries (South Africa, Kenya, Morocco, Ghana and Egypt) they were selected as potential feedstocks for the biochemical methane potential (BMP) tests performed in the Biowaste4SP project. Table 2.6 shows the ultimate analysis results as well as P, K, and Na content of the samples.

Table 2.6: Ultimate analysis for selected biowaste residues

Biowaste	C (%)	H (%)	N (%)	O (%)	S (%)	P (g/100g)	K (g/100g)	Na (g/100g)
Sugar cane bagasse	44.4	5.72	0.18	33.7	0.00	0.89	0.39	0.390
Fruit & vegetable waste	40.5	5.75	2.87	30.1	2.05	20.0	19.2	0.107
Banana (whole fruit)	40.5	5.56	0.91	31.1	0.00	20.3	64.4	0.018
Cattle manure	22.5	3.29	1.38	14.9	1.86	32.3	14.7	0.149

From the results shown in Table 2.6, sugar cane bagasse, vegetable waste and banana (whole fruit) were highly carbonaceous (C > 40%). The hydrogen content of all four biowaste residues was within the same range (3 to 6%) and the nitrogen contents of all the four biowaste materials were very low (0 to 3%). However, sugar cane bagasse had the lowest nitrogen content (0.18%) among the four biowastes: this is a disadvantage if sugar cane bagasse is to be used in anaerobic digesters (AD) and composting processes. The carbon : nitrogen (C:N) ratio for sugar cane bagasse in this study was calculated to be 246.6:1 (Table 2.7). This is too high a C:N ratio for anaerobic digestion and composting processes, where optimum ratios are required to be in the range 20:1 to 40:1 (Alvarez, 2003), (Silva & Naik, 2006). Sugar cane bagasse and banana (whole fruit) contained no sulphur, whereas the sulphur content for vegetable waste and cattle manure was very low (approximately 1 to 2%). Sulphur is an undesirable pollutant in most processing technologies due to formation of hydrogen sulphate (H_2S) in anaerobic digestion that reduces the quality and calorific value of biogas (Monnet, 2003). Sugar cane bagasse had a considerably lower nutrient content, with respect to P and K, compared to the other materials. However, it had the highest Na content (0.39g/100g), an element usually classified as a trace element (Gerardi, 2003).

Table 2.7: C:N ratios for selected biowaste residues

Parameter	Sugar cane bagasse	Vegetable waste	Cattle manure	Banana (whole fruit)
C/N	246.6	14.1	16.3	44.5

CONCLUSIONS

This study dealt with the identification and characterisation of solid biowaste residues in selected areas in South Africa with the objective of valorising them into high value products to avoid landfilling them. Twelve biowaste residues were considered and analysed. Raw water hyacinth had the highest moisture content as indicated by the weight loss during sample preparation. This means that processing water hyacinth to produce valuable products would require substantial water removal; hence, extensive pretreatment would be warranted. Results obtained from dry, and ash content determination indicates that all the identified biowastes contained volatile solids greater than 60%. Sawdust had the highest VS content (93%) whereas soybean waste and water hyacinth had the lowest VS content (63%). Analysis of carbohydrate content indicated that glucan is the most abundant carbohydrate in the identified biowaste residues. Banana (fruit only) had a higher nutrient content (protein, starch and fats) compared to the other bioresidues identified. This implies that banana could be a potential feedstock, especially for biogas production. Elemental analysis results revealed that sugar cane bagasse had a lower nutrient content, as evidenced by the significantly higher C:N ratio (246.6) as well as lower P and K content, compared to the other biowastes residues implying that it is unattractive for biogas production.

ACKNOWLEDGEMENTS

The authors hereby acknowledge the European Union for funding the Biowaste4SP project. This work was conducted as part of the international Biowaste4SP project, involving five participating African countries and support institutions from Asia and Europe (see www.biowaste4sp.eu). A sincere thank you to the eThekwini Municipality for co-funding and supporting the research, and to the CSIR and the University of KwaZulu-Natal for the use of their laboratory facilities.

REFERENCES

Alvarez JM. (ed.). 2003. *Biomethanization of the organic fraction of municipal solid wastes*. IWA Publishing, London.

Barletta D, Berry RJ, Larsson SH, Lestander TA, Poletto M and Ramirez-Gomez A. 2015. Assessment on bulk solids best practice techniques for flow characterization and storage/handling equipment design for biomass materials of different classes. *Fuel Processing Technology*, 138, 540—554.

Baserga U. 1998. *Landwirtschaftliche Co-Vergärungs-Biogasanlagen*. FAT-Berichte Nr. 512, Eidg. Forschungsanstalt fur Agrarwirtschaft und Landtechnik, Tänikon, Schweiz (Agricultural co-fermentation, biogas plans. FAT-report no. 512, Swiss Federal Research Station for Agricultural Economics and Agricultural Technology).

Department of Environmental Affairs (DEA). 2012. *National Waste Information Baseline Report*-Draft 6. Department of Environmental Affairs, Pretoria, South Africa.

Eaton AD, Clesceri LS, Greenberg AE and Franson MAH. 1998. *Standard Methods for the Examination of Water and Wastewater*. 20th ed. American Public Health Association, American Water Works Association, Water Environment Federation (WEF). Washington, DC.

Eliyan, C. 2007. *Anaerobic Digestion of Municipal Solid Waste in Thermophilic Continuous Operation*. Asian Institute of Technology, School of Environment, Resource and Development, MSc Thesis. Thailand, pp. 10–45.

Gerardi, MH. 2003. *The Microbiology of Anaerobic Digesters*. John Wiley & Sons. New Jersey, pp. 11–25.

Gustavsson, M, Bjerre, A, Bayitse, R, Belmakki, M, Gidamis, A B, Hou, X, Houssine, B, Owis, AS, Sila, DN, Rashamuse, K, Sundquvist, JO, El-Tahlawy, Y and Tawona N. 2014. *Catalogue of Biowastes & Bioresidues in Africa*. Deliverable Reports 1.3 & 2.3 (Biowaste4SP project). Available at: www.biowaste4sp.eu. [Accessed 12 August 2015].

Lestander TA. 2010. *Biomass characterisation by NIR techniques*. Biomass Technology and Chemistry, Swedish University of Agricultural Sciences (SLU). PowerPoint Presentation, Vienna, Austria. Available at: www.abo.fi. [Accessed 15 July 2015].

Luque, R, Herrero-Davila, L, Campelo, JM, Clark, JH, Hidalgo, JM, Luna, D, Marinas, JM and Romero, AA. 2008. Biofuels: a technological perspective. *Energy & Environmental Science*, 1(5), 542–564.

Monnet, F. 2003. *An introduction to anaerobic digestion of organic wastes*. Remade Scotland, Report. Scotland, pp. 48.

Oelofse, S and Muswema, AP. 2013. South Africa country report: Overview of potential biowaste and biobased residues for production of value added products. M. Gustavsson, Report. Council for Scientific and Industrial Research (CSIR), South Africa – Biowaste4SP project, European Union – 7th Framework Program. 39. Publ. No: CSIR/NRE/GES/ER/2013/0050/A.

Pradhan, A and Mbohwa C. 2014. Development of biofuels in South Africa: Challenges and opportunities. *Renewable and Sustainable Energy Reviews*, (39), 1089–1100.

Raclavska, H, Juchelkova D, Roubicek V and Matysek D. 2011. Energy utilisation of biowaste – Sunflowerseed hulls for co-firing with coal. *Fuel Processing Technology*, 92(1), 13–20.

Silva, MRQ and Naik, TR. (2006. Review of composting and anaerobic digestion of municipal solid waste and a methodological proposal for a mid-size city. In: Chun, YM., Claisse, P., Naik, TR, and Ganjian, E. *Sustainable Construction Materials and Technologies*. Taylor & Francis Group, London, pp. 631–644.

Sundqvist, JO, Gustavsson M, Bartali EH and Belmakki M. 2013. Protocol for the selection and storage of biowaste feedstock in the Biowaste4SP project. IVL Swedish Environmental Research Institute and Institut Agronomique et Vétérinaire Hassan II – Report, Biowaste4SP project. Available at: www. biowaste4sp.eu.

Taylor, G. 2008. Biofuels and the biorefinery concept. *Energy Policy*, 36(12), 4406–4409.

Theander, O, Aman, P, Westerlund, E, Andersson, R and Pettersson, D. 1995. Total dietary fiber determined as neutral sugar residues, uronic acid residues, and Klason lignin (the Uppsala method): collaborative study. *Journal of the Association of Official Analytical Chemists International*, 78(4), 1030–1044.

United Nations Environment Programme (UNEP) 2012. *UNEP Annual Report*. United Nations. Available at: http://www.unep.org. [Accessed 13 June 2015].

Zhijia L, Bingbing M, Zehui J, Benhua F, Zhiyong C and Xing'e L. 2016. Improved bulk density of bamboo pellets as biomass for energy production. *Renewable Energy*, 86, 1–7.

BIOLOGICAL TREATMENT

EVALUATION OF THE APPLICABILITY OF DRAFT NATIONAL NORMS AND STANDARDS FOR ORGANIC WASTE COMPOSTING TO COMPOSTING FACILITIES ON LANDFILL SITES

R du Plessis

University of South Africa, Department of Environmental Science,
P.O. Box 392, Pretoria, 0003, South Africa

Corresponding author e-mail: dplesr@unisa.ac.za

ABSTRACT

The reduction in the generation and disposal of waste has become an urgent priority in South Africa as evidenced in the promulgated National Environmental Management: Waste Act (RSA, 2008). In South Africa, the most common waste management method is by landfill. Most of these landfills do not comply with the standards for modern landfills, resulting in a negative impact on the environment. Approximately one third of all municipal waste landfilled is organic waste, which is biodegradable and can be converted to compost. Based on a feasibility assessment of establishing composting facilities on municipal landfill sites, this study applied the Draft Norms and Standards for Organic Waste Composting to facilities using a set of criteria. The draft Norms and Standards for Organic Waste Composting were developed to give effect to the National Organic Waste Composting Strategy aimed at beneficiating organic waste by promoting composting as one treatment option. The results show that the criteria are generally compliant with the suggested norms and standards. The areas that are lacking are of a legal nature, specifically with regard to monitoring of the operations. It was found that the draft Norms and Standards for Organic Waste Composting address legal, environmental, social and operational issues to protect the environment and the surrounding landowners.

Keywords: composting, organic waste, waste minimisation, landfill sites, criteria for composting facilities

INTRODUCTION

Background

In South Africa, landfill sites are under pressure due to the increased volumes of waste generated by a growing population and economy (DEA, 2012a). The National Waste

Management Strategy (NWMS) was developed to improve waste management and specifically promote waste minimisation in all its forms (DEA, 2012a). The green waste fraction of the municipal waste disposed of in landfill sites has a cumulative negative effect on the environment due to the high nutrient content and the ability to decompose rapidly, in the process releasing harmful greenhouse gases and leachate (Botkin & Keller, 2007; City of Illinois Extension, 2016; Diaz *et al.*, 2007).

Organic waste is a renewable resource which can be converted into compost, a humus-like substance in the pre-treatment of municipal solid waste (MSW), to produce a soil amendment that can enhance physical, chemical and biological soil properties (Diaz *et al.,* 1993; Pan *et al.,* 2012).

Policy environment

In South Africa, the responsibility for managing MSW falls to local governments, as assigned by the Constitution of South Africa (RSA, 1996). The National Environmental Management: Waste Act (Section 11(4)) (RSA, 2009), requires local municipalities to develop integrated waste management plans, which, after approval by the Member of the Executive Council (MEC), must be incorporated into the Municipal Integrated Development Plan as contemplated in Chapter 5 of the Municipal Systems Act (RSA, 2000). Municipalities are therefore legally compelled to manage the waste generated within their boundaries according to legislation and have the authority to determine the method of disposal. The need for application of the waste hierarchy, including waste minimisation, reuse and recycling, has become essential to reduce waste disposal to landfill sites in an attempt to prolong the lifespan of the existing landfill sites which are rapidly filling up (CSIR, 2011; DEA, 2012a). To give effect to the National Waste Management Strategy (NWMS) (DEA, 2012a), the Department of Environmental Affairs (DEA) developed the National Organic Waste Composting Strategy (NOWCS) (DEA, 2013a) specifically aimed at addressing the green fraction of MSW.

The aim of this study is to present a solution to waste minimisation through the establishment of composting facilities, in line with current legislation, on landfill sites. There is no evidence in literature that this is being done either nationally or internationally.

National Organic Waste Composting Strategy

In an attempt to regulate general waste, section 20 of the Environmental Conservation Act (ECA), (RSA, 1989) was applicable for the permit application to run a composting operation. Thereafter, the Environmental Impact Assessment (EIA) regulations of 2006 (GN 386, 1(o)) regulating handling of general waste of more than 20 tonnes per day. This regulation was repealed and replaced by Category A of GN 718, issued on 3 July 2009 in terms of the National Environmental Management: Waste Act (RSA, 2008) and required that a basic assessment report be submitted for review with the aim to obtain a waste management licence (WML) (DEA, 2009). Currently, GN 921 of 29 November 2013 is the applicable legislation for the application to treat general waste, therefore also green waste (DEA, 2013b).

To promote the establishment of composting facilities and thus prevent green waste being landfilled and to alleviate legislative requirements of applying for a WML, the DEA initiated the compilation of a strategy for composting of green waste (DEA, 2013a): "*This project is to strategise the potential of composting as a method to beneficiate organic waste, as one of a basket of options to help divert organics from landfill disposal*". The National Environmental Management: Waste Act: National Norms and Standards for Organic Waste Composting (draft report) (hereafter referred to as N&S) was consequently published on 7 February 2013, suggesting minimum requirements for the operation of a composting facility and is currently awaiting promulgation.

The limit is set for a facility with the capacity to process in excess of 10 tonnes but less than 100 tonnes of compostable organic waste per day. The same limit as in listed activity A(6) of GN 921 is used, which must be adapted (to exclude organic waste) before the N&S for composting can be effected (Gower-Jackson, 2015). In terms of composting facilities on landfill sites, no additional legal requirements other than compliance with the N&S will be discussed.

Municipal response

It is common practice to have drop-off points for recyclables, including green waste at garden centres (also known as transfer stations) from where the waste is transported to composting facilities (Furter, 2004; Pikitup, 2007).

Since April 2001, Cape Town has diverted the green waste from landfill sites by creating collection points at conveniently located garden sites (Furter, 2004). At 11 of the 17 council drop-off facilities, the incoming garden waste is collected and the volume is reduced to 25% through chipping. Through private partnerships, the green waste is collected from various facilities for composting, thereby creating savings for the municipality in terms of reduced transport costs and landfill airspace (Furter, 2004). The City of Tshwane Metropolitan Municipality (CTMM) has 10 garden sites for the convenience of the residents. From there, green waste is transported to the landfill site by the municipality (CTMM, 2015). The City of Johannesburg Metropolitan Municipality (CJMM) has 48 garden drop-off sites, one of which, the Panorama composting site in Roodepoort, received most of the green waste collected from the garden sites and converted it into compost, between 1994 and 2015 (Pikitup, 2016).

Composting facilities on landfills

The NOWCS (DEA, 2013a) makes it clear that composting is regarded as a non-viable business and support from local government is needed. To reduce transport costs, a composting plant should be located near to the source of the waste, implying more expensive land within municipal borders. A potential compost operator will not be able to buy or lease expensive land and run an economically viable operation. Existing landfill sites within the city are therefore attractive land options for the establishment of composting facilities. An evaluation in this regard was done to determine the viability of four sites in the CTMM (Du Plessis, 2008, 2010). The municipality owns the landfill sites and it makes economic sense to use these sites for monetary gain, either by selling the compost or using it on municipal parks, or by leasing out the land. These composting facilities may be able to support small business development and entrepreneurship in partnership with the municipality. The NOWCS promotes poverty alleviation and promotion of employment, which will be possible on all economic levels ranging from the developer of the composting plant to service providers and small entrepreneurs collecting and delivering organic waste.

The aftercare of a landfill site is prescribed as a condition in the WML. The study by Misgav *et al.* (2001) addresses the worldwide problem of aftercare for a landfill site, namely the reuse of sanitary landfills as an asset and a source of income. Landfill sites will therefore be the ideal location for the siting of composting facilities because the environmental, operational and social aspects have already been dealt with.

PURPOSE AND METHODOLOGY

Criteria for evaluating composting facilities on landfills

A case study, focusing on the Panorama landfill site in the City of Johannesburg, established criteria to determine the requirements for a composting facility specifically located on a landfill site (Du Plessis, 2010). These criteria were based on various strategic national and provincial government documents including the Minimum Requirements for Waste Disposal by Landfill (DWAF, 1998), which was the guiding document for issuing permits for waste disposal at the time the composting facility was established in 1999; the National Environmental Impact Assessment (EIA) regulations (RSA, 2006), which was used for the application process for a waste management licence and the guiding regulations during the time of the study in 2010; and the final version of the draft General Waste Management Facilities Standards of June 2009, which classified facilities according to operational requirements and the effect the facility would have on the environment (GDACE, 2009). The resultant criteria identified for the assessment of a composting facility on a landfill site are depicted schematically in Figure 3.1, and included physical, operational and social criteria.

Evaluation of criteria

In this study, the various physical, operational and social criteria for the establishment of a composting facility on a landfill site (hereafter referred to as "criteria") (Figure 3.1) are evaluated according to the latest suggested national legislation which is in line with the purpose of the NEMWA: Draft Norms and Standards for Organic Waste Composting. Although not yet finalised, the N&S are used as a yardstick to evaluate the "criteria" to determine their applicability. It was considered necessary to apply the N&S before they are promulgated to ensure that weaknesses, should there be any, are identified and corrected. This had not been done in any study and may provide useful insight into the suggested N&S.

This study therefore combines the NOWCS with three of the goals contained in the waste hierarchy, namely improving the management of landfilling through the separation of green waste, reducing waste to landfill by diverting the green waste from being landfilled, and recycling the green waste through composting.

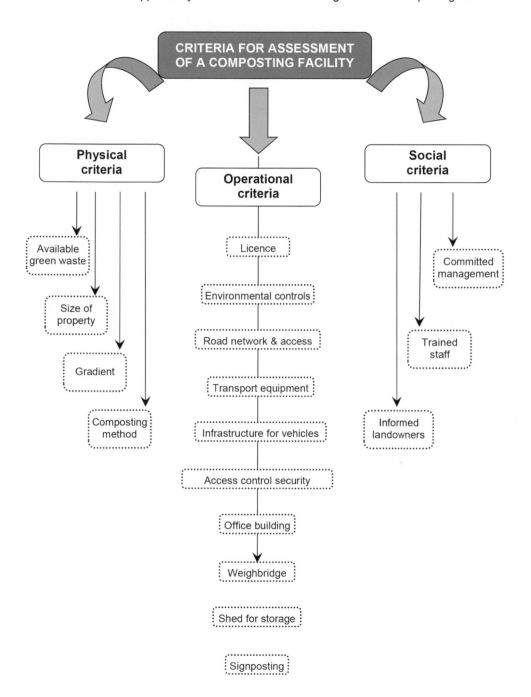

Figure 1:   Schematic representation of criteria for evaluation of composting facilities on landfill (adapted from Du Plessis, 2010)

MINIMUM REQUIREMENTS FOR THE ESTABLISHMENT OF COMPOSTING FACILITIES

The following section examines each of the physical, operational and social criteria depicted in Figure 3.1 with regard to their applicability should a composting facility be established on a landfill site; following which, the N&S in each case will be briefly discussed (Du Plessis, 2010).

Physical criteria

Municipal resources

Two resources, which lie within the boundaries and under the jurisdiction of a municipality, should be considered when establishing a composting facility on a landfill site (i) the green fraction of the general waste and (ii) available space on a landfill.

Green waste

Green waste, as a resource to produce compost, is available within the boundaries of local municipalities and could be used for composting, provided there is enough green waste available from municipal gardens and residential areas to make it sustainable. This availability varies from region to region depending on climatic area and waste composition (Taiwo, 2011). In India, the organic fraction of the solid waste is estimated at 40% to 85% (Zurbrügg *et al.*, 2004). According to Williams (1998), the compostable content of the waste stream in the United Kingdom can be as high as 60% if paper, cardboard and putrescibles are included.

Green waste in Johannesburg comprises 30% to 45% of the municipal waste stream (Pikitup, 2007). Snyman (2009) conducted a waste characterisation of the domestic waste in the City of Tshwane Metropolitan Municipality and found that in 2001, 28% and in 2004, 32% of the domestic waste sent to landfill was green waste. The National Waste Information Baseline Report states that the organic waste composition as measured by mass for Tshwane is 15% and by source 32%; for Cape Town, as measured by mass is 18% and by source 5%; Johannesburg as measured by source is 6% and for Mangaung it is 17%. The National Waste Information Baseline Report concluded that general waste generated in municipalities is 40% in total of which 13% is organic waste (DEA, 2012b) (32.5% of general waste is green waste). Green waste is classified as general waste (GW20) (DEA, 2012b) and falls under the same regulations.

Landfill space

Available space on a landfill can be regarded as a non-renewable resource due to the high cost of constructing a sanitary engineered landfill site, high land prices and limited land availability in the vicinity of the waste generators to save on rising transport costs (Chang & Davila, 2006). The NIMBY ("not in my backyard") syndrome creates difficulties with siting new landfill sites, for example, windblown litter creates a visual impact, leachate pollutes groundwater, noise is generated by waste trucks and compactors, bad odours, and dust can cause health problems (Williams, 2005).

Regarding relevance in terms of the N&S, paragraph 5(3) states non-conforming waste must not be accepted at the site. When it comes to establishing a composting facility on a landfill, issues of gradient and available space need to be considered.

Gradient

The site must be sloped to prevent ponding. The required slope of 2% to 4% (EPA, 1994) will be graded during the construction phase of the composting facility, and needs to have a site-specific design by a professional engineer, including an impermeable surface to protect the soil and groundwater from possible contamination.

This is addressed by the N&S in paragraph 4(3) requiring the construction of drainage systems for surface run-off wastewater; 4(4) that a single composite line is needed; 4(5) run-off must be diverted from the site; 4(9) preventing the run-off to come into contact with the incoming stock and materials being processed; 6(d) leachate must not contaminate storm water run-off; and 5(e) contaminated run-off may be sprayed over the compost.

Composting method

The method of composting used is dependent on the available space at the landfill site. Smaller areas will not be able to accommodate the most common windrow method but will, for example, need to use in-vessel composting (EPA, 2015). In-vessel composting will also be preferable if the buffer zone of the landfill site has been compromised by encroaching residential or commercial land use, because the odour is easier to contain (EPA, 2015).

The N&S do not prescribe the composting method but provides operational guidelines in paragraph 6 regarding air emissions control, vermin and leachate.

Operational criteria

Licence

Once the N&S are promulgated, the licensing process will be replaced by an application process to register the composting facility with the competent authority should the volumes fall within the limits set by the N&S: treatment of organic waste of more than 10 tonnes but less than 100 tonnes per day as stipulated by paragraph 3 of the N&S. This will be a separate process from the legal requirements for a landfill site (DEA, 2013b).

Environmental controls

The treatment process of the green waste can have certain adverse effects affecting the natural, social and economic environment around it (EPA, 2015; Williams, 2004).

Air quality

Bad odours, the creation of dust and bioaerosols can affect air quality as a result of composting activities (DWAF, 1998). Air quality can have an impact on human health and is therefore regulated by setting acceptable air emission standards for South Africa in the National Environment Management: Air Quality Act (Act No. 39 of 2004)

(NEMAQA). The Occupational Health and Safety Amendment Act (Act No. 181 of 1993) also regulates health issues (RSA, 2004). The requirements of this legislation needs to be taken into consideration when the method of composting, as well as operational guidelines, are selected to prevent degradation of the air quality in the immediate vicinity of the composting facility. During composting, a natural breakdown of organic material produces primarily carbon dioxide and water vapour, and heating takes place. When this process is unbalanced, other gases with objectionable odours may be released (Composting Council of Canada, 2006). Odour management is therefore one of the key reasons why the composting process needs to be optimised, choosing the most suitable method.

Gases, which may be generated and hence managed in the composting process, include carbon dioxide (CO_2), methane (CH_4), hydrogen cyanide (HCN), and hydrogen sulphate (H_2S).

Dust may be generated from dry, uncontained organic materials. This will increase during screening and shredding, and from vehicles driving over unkempt roads. Dust can clog equipment and may carry bacteria and fungi that can affect the health of workers at the facility (EPA, 1994). Dust is increased during the windrow turning process. Therefore, it is important not to mix the compost on windy days and to water the windrow piles while turning, thereby reducing the amount of dust.

Bioaerosols are suspensions of particles in the air consisting of micro-organisms that can be inhaled. A very commonly found fungus is *Asperigillus fumigates.* Although, it does not pose a health hazard to healthy people, it might affect susceptible persons, and can bring about a weakened immune system, asthma, diabetes and allergies, to name but a few. *Asperigillus fumigates* is often found in the dust of decaying organic material, however, concentration levels decrease rapidly over a short distance (Smith, 1994). An appropriate buffer zone, as well as dust masks for workers are appropriate measures to minimise the effect of bioaerosols.

Mitigation measures can reduce the effect of odours to improve air quality. The composting method and the carbon/nutrient ratio will play an important role in the amount of odours and emissions that will be released (Phillips, 2002). This will also determine the type of material that can be accepted for composting. Should odorous material arrive, it is best to mix it immediately into high-carbon material. A layer of finished compost can be used to cover the outside of the pile to act as a biofilter if windrows are used. Local weather conditions need to be taken into account before the compost is turned or moved. Wind direction and speed will determine dispersion of potential odorous or dusty materials to neighbouring areas. It is not recommended that windrows be turned on a windy day. A low barometric pressure may cause gases to flow at ground level, and high pressure can cause it to disperse. It is necessary to ensure that there is good drainage (Phillips, 2002). Lastly, the composting facilities with offensive odours can be located in an area where the effect on daily lives will be minimal (EPA, 1994), for example, a buffer zone to protect residents or occupants of neighbouring properties. Additionally, personal protective equipment (PPE), such as facemasks for the workers, will minimise the effect of possible harmful bioaerosols.

Air pollution is not a major problem if the composting facility is managed properly. Finstein emphasised the fact that no composting facility will survive politically if the odour control is not effective (Montague, 1993). Large areas are therefore needed to minimise the effect of bad odours, dust and bioaerosols, rendering a landfill site with an accompanying buffer, a suitable option. Alternatively, should a buffer zone not be available, an in-vessel composting method can be implemented. This is, however, far more expensive than open windrows.

The N&S address air quality in paragraph 6(1) by advising on (a) minimisation of odour emissions, (b) covering rapidly biodegradable organics, (c) to prevent dust by watering the roads, (f) to keep the organics moist to minimise emissions of airborne pathogens and (g) and (h) to aerate windrows.

Fires

High temperatures of up to 93°C in dry compost piles can lead to spontaneous combustion. This is only possible when windrows are four meters and higher. The risk of fires can be reduced by keeping the windrows to three meters or lower, employing good site security to prevent arson and dumping of flammable materials such as diesel, as well as the prevention of the accumulation of dust (EPA, 1994).

The Minimum Requirements (DWAF, 1998, 2005) that were used as a guideline for the waste management licences of most of the established landfill sites make provision for good site security on a landfill site, including, *inter alia*, fire-fighting equipment, emergency numbers and access to the site for emergency vehicles, such as fire-fighting vehicles. Spontaneous fires are common on active landfill sites. Therefore, it is suggested that composting facilities be established either on old cells or in open, unused areas to prevent the possibility of spontaneous fires.

Paragraph 4(2) of the N&S requires that the composting facility be accessible for emergency vehicles. Paragraph 6(4) specifies fire and methane gas management, and paragraph 5 addresses the need for site security and access control at a composting site.

Water management

On-site water management must be considered in terms of surface water (run-off or storm water), groundwater and leachate. If storm water is not controlled properly, flooding can create the risk of erosion and can wash away the windrows. Standing water will create muddy conditions and consequently water pollution as well as increasing operational costs, since the windrow turner will not be able to access the site (EPA, 1994). Windrows should be constructed along the slope of the land instead of across it to encourage effective drainage (Phillips, 2002).

Water channels, cut-off berms, drainage ditches, or interceptor drains should be implemented to control storm water (Phillips 2002). Rainwater on the windrow area should be channelled to an evaporation pond or treatment facility and can be used to spray the windrows (DEA, 2014). The storm water channels must be maintained to ensure efficient operation. Composting facilities should not be allowed in areas located within the 1:50 and 1:100 year flood lines (DWAF, 1998).

With reference to composting, Phillips (2002b) describes leachate as "*liquid that has percolated through and drained from feedstock or compost and has extracted dissolved or suspended materials*". The main concerns of leachate are the enhanced nutrient load, biochemical oxygen demand (BOD), and the presence of phenols. A high BOD can be a potential threat to aquatic life because it depletes the dissolved oxygen in surface water bodies (EPA, 1994).

A common source of nutrients in composting piles is grass clippings (EPA, 1994). Adding carbon to keep the carbon nutrient ratio of the compost in balance will help minimise the loss of nitrogen into leachate (Phillips, 2002). When excessive quantities of nitrate are released into water sources, it can lead to eutrophication, and a sudden increase in algae, which depletes the oxygen so that bacteria in the water die, lowering the dissolved oxygen, which causes fish to die (Botkin & Keller, 2005). Therefore, wetlands, floodplains, surface water, and groundwater should be protected from leachate (EPA, 1994).

The advantage of having a composting facility on a landfill site is that storm water control measures, in the form of a cut-off trench, are a requirement of the licence conditions when the landfill site is established. It is a large capital layout, and could therefore save initial costs during the establishment of a composting facility.

The N&S paragraph 6(6) advises on water pollution prevention: (a) to prevent surface water to come into contact with the organics on site; (b) to operate on a protection slab with drains leading into a sump; (c) to use sump water for watering the windrows and (d) to treat leachate on site.

Noise

In terms of Regulation 7 of the Environmental Regulations for Workplaces, promulgated under section 35 of the Machinery and Occupational Safety Act (Act No. 6 of 1983) (RSA, 1983), no employer may require or permit an employee to work in an environment in which he or she is exposed to a noise level equal to or exceeding 85 decibels (South Africa: Consolidated Regulations, 2003). To put this into context, the Gauteng Noise Regulations suggest 70 decibels for industrial use and 55.5 decibels for residential areas (South Africa: Consolidated Regulations, 2003). Noise at a composting facility is generated by incoming and outgoing trucks, and equipment such as hammer mills, shredders and grinders, turners and front-end loaders. The noisiest of this equipment is about 90 decibels at the source (EPA, 1994).

Noise reduction measures include:

- operation of the composting facility must be restricted to normal business hours to protect surrounding neighbours;
- equipment must have noise reduction features such as noise hoods and mufflers;
- noise generation and noise reduction equipment must be properly maintained;
- operators of the equipment, as well as those working in close proximity of the sound generating equipment, must be provided with hearing protection as prescribed by the Health and Safety regulations; and
- a wall around the composting facility, a soil berm, trees or a buffer zone will provide effective reduction in the sound for the surrounding communities (EPA, 1994).

The required buffer zone around a landfill site will reduce the noise problem for surrounding landowners, should a composting facility be established there.

Paragraph 4(6) of the N&S addresses the reduction of noise levels during construction but not during operation, which is regulated by local noise regulations and therefore covered in terms of legislation.

Vermin and disease vectors

A composting facility, as with any waste management facility, has the potential to attract pests (Composting Council of Canada, 2006). Mice, rats, mosquitoes and flies – all potential carriers of diseases – may be attracted by the food and shelter available at composting facilities. This may cause health hazards and needs to be controlled by proper operating procedures (EPA, 1994) regulating the correct method of composting. Offensive odours are a sign that there is a problem with the composting process and indicate an increased risk of attracting vermin. The most common causes of such odours are inadequate aeration or excessive moisture. Prevention of bad odours, as well as keeping the site clean, will ensure that the site stays free of rodents and other pests (Composting Council of Canada, 2006). Turned windrows, with regular turning, are therefore the best option. Static windrows, which are not turned, should be covered with a layer of mature compost to prevent odour and vermin (EPA, 2015).

Rodents can be controlled by keeping cats or owls on the premises. A professional exterminator may be contracted if the problem becomes uncontrollable. The risk of houseflies, which can transmit salmonella and other food-borne diseases, is minimised with the high temperatures reached in the composting process (which can kill all life stages of the housefly). Mosquitoes breed in standing water; therefore, ponding of water must be prevented. The transmission of diseases may be prevented through good housekeeping, maintaining proper aerobic conditions and high temperatures (up to 57°C) and appropriate grading of the land (EPA, 1994).

Most of the requirements above should be in place at a landfill site. The requirement that the waste needs to be covered on the landfill at the end of each day prevents odours and decreases the presence of flies and rodents on the premises (DWAF, 1998). Although composting will not be covered with soil daily, the principle of good housekeeping, as will be prescribed by an environmental management programme as part of the WML of the landfill, is required.

The N&S paragraph 6(1)(b) stipulates that food waste be contained in vermin proof containers and paragraph 6(5)(a) advises on keeping vermin populations as low as possible.

Litter

As green waste is part of the general domestic waste stream, provision must be made to accommodate litter. Windblown litter may be caused by incoming yard trimmings and must be managed by using permanent or movable fences to facilitate the collection of the litter. Incoming vehicles must be covered to prevent litter from blowing out. Litter must be contained and cleared away as soon as it occurs, before it spreads off-site (EPA, 1994).

Should the composting facility be sited within the premises of a landfill site, the problem of litter from the composting facility will be minor in comparison to that of the landfill operations, if landfilling and composting coexist. The composting facility will be contained to a specific area and waste management must be enforced to keep the site tidy from litter (Goedhart, 2010). This is usually one of the licencing conditions.

The N&S paragraph 6(2) (v) stipulates that windblown litter must be contained.

Health, safety and security

Exposure to bioaerosols, other potential toxic substances, excessive noise and injuries from equipment, can all pose health and safety threats (EPA, 1994). These can be prevented by proper operation of the facility, adequate training, adequate site security and compliance with provisions of the Occupational Health and Safety Act 85 of 1993 (South Africa: Consolidated Regulations, 2003). Landfill sites are monitored regarding health, safety and security, which could be extended to the composting facility.

Security is addressed in paragraph 5 of the N&S and Health and Safety measures are contained in most of the specifications of the N&S for the operations of the site to ensure healthy and safe working conditions.

Visual

The fact that composting facilities should be accessible and close to the generation of waste, thereby reducing transport costs and stimulating participation in local composting programmes, necessitates measures to screen off the facility to reduce the visual impact. This can be done by the construction of artificial buffer zones, for example, earth berms, trees or walls (EPA, 1994). Should a composting facility be established on a landfill site, the visual impact would not be an important issue to consider, since large machinery, as well as mountains of waste and landfilling operations are already present on the site.

Visual aspects are addressed by the N&S under paragraphs 6(2)(v), (3)(a), (5)(c) and possibly 11(2).

Road network and access

The ideal is to have land available as close to the source of green waste as possible. Open land within the borders of a city is very expensive and scarce, and it is unlikely that prime land will be used for composting (Du Plessis, 2008).

The composting area must be fenced off and access control must regulate traffic during construction and operation phases according to N&S paragraph 4(8) and paragraph 5(1).

Transport equipment

Transport, usually in the form of large trucks, is needed to deliver the incoming waste stream to the composting site. To reduce traffic at the facility, the same vehicles could be used to transport the compost to the user.

No specifications regarding transport are provided in the N&S other than that no maintenance may be done on site (paragraph 4(7)) and that vehicles leaving the site must not leave mud tracks (paragraph 6(5)(b)).

Infrastructure for vehicles

Signposts, speed limit measures (speed humps) and direction signs must be clearly visible. The access road must be an all-weather road, suitable for wet and dry conditions (Du Plessis, 2006). This road needs to lead directly to the composting facility to drop off green waste or collect compost for distribution. Municipal landfill facilities will have a constant flow of traffic; therefore, a dedicated road needs to lead to the tipping and storage area (EPA, 1994).

The only reference in the N&S regulating the installation of underground fuel storage tanks for use by vehicles is paragraph 7(8).

Access control security

Site security requires access control to the site to prevent theft, vandalism, arson or other offences. A security fence around the site needs to be in place (Goedhart, 2010). Access control security is likely to be in place should a composting facility be established on a landfill site.

Paragraph 5 of the N&S recommends access control to prevent unauthorised entrance and screening of waste before it enters the site to ensure that only permissible waste enters.

Office building

Composting is a business and office facilities are required for this purpose. As a minimum standard, electricity, drinking water, telephone connection and toilet facilities should be available, even for small operations (Goedhart, 2010). Depending on the size of the office of the landfill site, it might be possible to share office space to save on costs.

Paragraph 4(8) of the N&S mentions that areas under construction be demarcated. No mention of office buildings in particular is made.

Weighbridge

For operational and economic purposes, it is essential to know how much waste enters and how much compost leaves the site. A weighbridge ensures accurate recordings. According to the National Waste Information Regulations, waste quantities must be reported on a central waste information centre (DEA, 2012c). This will apply for both the landfill site and the composting facility, should it be established on the same premises. Accurate recordings can only be done with a weighbridge, which can be utilised by the composting plant, to save on a major expense.

N&S paragraph 6(3)(d) suggests that incoming organics and outgoing compost be recorded. This record must be safely kept (paragraph 7(2)) and made available if required (paragraph 10(3)). The information must be reported annually on the waste information system applicable to that facility (DEA, 2012c).

Shed for storage

For economic considerations, it is important to protect the finished product from the elements. Should the composting facility be sited on a landfill site, a shed should be designated for the sole use of the composting facility.

Paragraph 4(3)(c) requires that an approved engineering drawing be used for the floor area design of the storage area. Paragraph 6(3)(c) stipulates that contamination of the final product be minimised.

Signposting

Signposts must clearly indicate the different operational areas, that is, where incoming waste must be offloaded. This will ensure the smooth running of the composting facility. This offloading point should not be shared with the landfill site, should the composting plant be on the same premises.

The N&S do not refer to signposting.

Social parameters

Committed management

This issue does not have a bearing on environmental issues other than the identification of a responsible person (applicant for a permit) to be held accountable for legal compliance. If the composting is a municipal function, the responsible person could be the same for both operations.

Paragraph 11(4) of the N&S mentions a responsible person when the site is decommissioned. This will also be covered by the registration process.

Trained staff

To ensure the effectiveness of the business as well as the health and safety of employees, staff need to be trained. Staff from the landfill site might already be trained regarding waste management aspects, however, ensuring that employees know how to work with specialised equipment and to handle other specialised tasks, will require applicable training.

Paragraph 8 of the N&S address training and capacity building.

Informed landowners

Due to the potential environmental and social impacts that a landfill site may have, the site is usually selected with great care and thorough investigation. The surrounding areas of landfill sites usually act as buffer zones to minimise the effect on the neighbouring communities (Du Plessis, 2008).

The N&S do not make specific reference to a public participation process; however, environmental controls to ensure minimum impact on the surrounding environment are specified.

EVALUATION OF NORMS AND STANDARDS FOR COMPOSTING FACILITIES ON LANDFILL SITES

From the above review of the previously established criteria for composting facilities on landfill sites, it is clear that the physical, operational and social criteria cover most of the minimum requirements stipulated in the draft Norms and Standards for Organic Waste Composting.

Specific minimum requirements in the N&S not matched to the "criteria" include –

- paragraph 4(1) siting the composting facility in a sensitive area;
- aspects of waste management, namely paragraph 6(2)(a)(i) and (iv), referring to liquid waste, 6(2)(a)(ii) requiring an integrated waste management plan, and 6(2)(a)(v) controlling hazardous waste;
- paragraph 6(3) refers to the handling of incoming and processed waste;
- regulating incoming waste to prevent it from exceeding the design requirements (paragraph 5(3)(a)); and
- the storage management of incoming raw organics (paragraph 6(3)(b)).

Shortcomings of the "criteria" used in this evaluation that were not already mentioned include section 7 of the N&S (General requirements) –

- promoting compliance with other legislation (paragraph 7(1));
- hazardous waste management (7(5)) (not applicable);
- an SDS for chemical products used (7(6)) (not applicable);
- handling of non-recyclable waste (7(7)); and
- registration with the DAFF (Department of Agriculture, Fisheries and Forestry).

Monitoring, auditing and reporting (paragraph 10) is also not covered in the above "criteria". However, this will also be a requirement of the Norms and Standards for the Storage of General and Hazardous waste, which will be the applicable legislation.

The result of this evaluation is that the physical, operational and social criteria developed in 2010, with amendments to include the few missing minimum requirements under the N&S, provide a useful tool for evaluating the siting of composting facilities on landfill site.

CONCLUSION

This chapter addressed the requirements of a composting facility specifically located on a landfill site to demonstrate the feasibility of such a facility. The relevance to the specifications required by the draft Norms and Standards for Organic Waste Composting (DEA, 2014) were highlighted.

Municipalities are under increasing pressure to divert waste away from landfills, as evidenced by many recent initiatives of government on waste minimisation. Although the organic waste content of MSW in South Africa differs between municipalities, the National Waste Information Baseline Report (DEA, 2012b) shows that the green waste fraction makes up approximately one third of the MSW stream in South Africa, most of which is still landfilled. This is in spite of a clear understanding of the environmental impacts associated with landfilling organic waste, as well as the fact that in the larger municipalities garden sites provide the opportunity to separate a large portion of the organic waste from the rest of the MSW stream.

It is essential that municipalities comply with legislation, provide a good service to its communities, and take responsibility for protecting the natural environment. It is therefore advisable that composting be strongly promoted as a technology solution for diverting organic waste from landfill. It is crucial to emphasise the importance of composting as a way to minimise waste, and to solve the growing waste problem by not continuing to waste green waste, which is a valuable resource. Additionally, the composting site must not be seen as an opportunity to profit from the compost, but rather it should be recognised for its value in diverting the green waste from landfills, saving landfill space and preventing the negative associated environmental impacts.

Establishing composting facilities on landfills, as an end-use for a closed landfill site, can be of economic value, which could also contribute to the maintenance of the site (DEAT, 1998). *"There are many challenges to re-using a closed landfill site. Liability considerations (toxic torts) and technical problems (settlement, gas, health and safety) abound. But just as a growing number of formerly-used industrial sites are being redeveloped for productive uses in what has become known as the "brownfield movement", so too have landfill sites been increasingly developed for high-value, productive land uses"* (McLaughlin, 2007).

REFERENCES

Botkin, DB and Keller, EA. 2007. Environmental Science: Earth as a living plane (3rd ed.). Danvers, IL: John Wiley.

Chang N-B and Davila, E. 2006. Minimax regret optimization analysis for a regional solid waste management system. *Waste Management*, 27(6), 820–832.

Chang, S, Cha, Y, Wu, M, Chiu, T, Che, J, Wang, F and Tseng, C. 2004. Acute Cycas seed poisoning in Taiwan. Clinical Toxicology, 42(1), 49–54. Available at: http://informahealthcare.com/doi/abs/10.1081/CLT-120028744 [Accessed 4 May 2010].

City of Illinois extension. 2016. Composting for the Homeowner. Available at: http://web.extension.illinois.edu/homecompost/science.cfm [Accessed 27 June 2016].

City of Tshwane. 2015. Waste Removal. Available at: http://www.tshwane.gov.za/sites/Departments/Agriculture-and-Environment-Managemental/WasteRemoval/Pages/Dumping_Sites.aspx [Accessed 24 May 2016].

Composting Council of Canada. 2006. Questions and answers about composting. Available at: http://www.compost.org//qna.html#section14 [Accessed 2 October 2015].

CSIR [Council for Scientific and Industrial Reasearch] 2011. Municipal waste management - good practices. Edition 1. CSIR, Pretoria.

DEA (Department of Environmental Affairs). 2009. GN 718. National Environmental Management: Waste Act 2008 (59/2008): List of waste management activities that have, or are likely to have, a detrimental effect on the environment. Pretoria: Government Printer.

DEA (Department of Environmental Affairs). 2012a. National Waste Management Strategy of South Africa. Pretoria: Government Printer.

DEA (Department of Environmental Affairs). 2012b. National Waste Information Baseline Report. Department of Environmental Affairs, Pretoria, South Africa.

DEA (Department of Environmental Affairs). 2012c. GN 625. National Environmental Management: Waste Act (59/2008): National Waste Information Regulations. Pretoria: Government Printer.

DEA (Department of Environmental Affairs). 2013a. National Organic Waste Composting Strategy (Draft document).

DEA (Department of Environmental Affairs). 2013b. GN 921. National Environmental Management: Waste Act 2008 (59/2008): List of waste management activities that have, or are likely to have, a detrimental effect on the environment. Pretoria: Government Printer.

DEA (Department of Environmental Affairs). 2014. GN 68. National Environmental Management: Waste Act 2008 (59/2008): Draft Norms and Standards for organic waste composting. Pretoria: Government Printer.

DEAT (Department of Environmental Affairs and Tourism). 2006. GN 386. National Environmental Management Act 1998 (107/1998): List of activities and competent authorities identified in terms of sections 24 and 24d of the National Environmental Management Act 107 of 1998. Pretoria: Government Printer.

Diaz, LF, Savage, GM, Eggebert, LL and Golueke CG. 1993. Composting and recycling municipal solid waste. London: Lewis Publishers.

Diaz, LF, Bertoldi, deM, Bidlingmaier, W and Stentiford, E. 2007. *Compost science and technology* (Waste Management Series, No. 8). Amsterdam: Elsevier.

Du Plessis R. 2008. Municipal composting facilities on selected sites in the Tshwane municipal area in South Africa. IWMSA 19th Waste Management Conference proceedings, WasteCon 2008, 6–10 October 2008. Durban: IWMSA.

Du Plessis, R. 2010. Establishment of composting facilities on landfill sites. MA Environmental Management dissertation: Department of Environmental Sciences, Unisa.

DWAF (Department of Water Affairs and Forestry), 1998. Minimum requirements for waste disposal by landfill (Waste Management Series, 2nd ed.). Pretoria: Government Printer.

DWAF (Department of Water Affairs and Forestry). 2005. Minimum requirements for waste disposal by landfill (Waste Management Series, 3rd ed.). Pretoria: Government Printer.

EPA (Environmental Protection Agency, US). 1994. Composting of yard trimmings and municipal solid waste. Available at: http://www.epa.gov/epawaste/conserve/rrr/composting/pubs/cytmsw.pdf. [Accessed 2 August 2015].

EPA (Environmental Protection Agency, US). 2015. Sustainable Management of Food. Available at: https://www.epa.gov/sustainable-management-food [Accessed 27 June 2016].

Furter L. 2004. Green waste should not end up in large holes in land. Resource, 6, 1. Available at: http://search.sabinet.co.za/images/ejour/resource/resource_v6_n1_a17.pdf [Accessed 10 October 2015].

GDACE (Gauteng Department of Agriculture, Conservation and Environment). 2009. General waste management facilities standards. Final version. Johannesburg: GDACE.

Goedhart, P. 2010. New Composting Facility, Builders Rubble Processing Plant and Domestic Drop-off Centre for George: Technical Report for Waste Management Licence.

Gower-Jackson, S. 2015. Personal communication via telephone. Pretoria. 15 September 2015.

McLaughlin, M. 2007. Closed landfills: assets in disguise. Available at: http://www.scsengineers.com/Papers/McLaughlin_Closed_Landfill-Assets_in_Disguise.pdf [Accessed 27 June 2016].

Misgav, A, Perl, N and Avnimelech, Y. 2001. Selecting a compatible open space use for a closed landfill site. *Landscape and Urban Planning*, 55(2), 95–111.

Montague, P. 1993. Rachel's Hazardous Waste News #352: News and resources for environmental justice. erf@igc.apc.org [Accessed 28 May 2010].

Pan, I, Dam, B and Sen, SK. 2012. Composting of common organic wastes using microbial inoculants. 3 Biotech. 2012 June; 2(2), 127–134. Published online 2011 November 2017. Available at: http://www.ncbi.nlm.nih.gov/pmc/articles/PMC3376866 [Accessed 27 June 2016].

Phillips C. 2002. The composting process: Leachate management. Composting Council of Canada. Available at: http://www.compost.org/pdf/sheet_3pdf [Accessed 26 June 2015].

Pikitup. 2007. Pikitup launches the first fully organic potting soil. Available at: http://www.pikitup.co.za/default.asp?id=860 [Accessed 20 October 2010].

Pikitup. 2016. Environmental XPRT. Available at: https://www.environmental-expert.com/services/commercial-services-268799 [Accessed 21 July 2016].

RSA (Republic of South Africa). 1983. Machinery and Occupational Safety Act 6 of 1983. GN R2281 Government Printers, Pretoria (16 October 1983).

RSA (Republic of South Africa). 1996. Act 108 of 1996. Constitution of the Republic of South Africa.

RSA (Republic of South Africa). 2000. Local Government: Municipal Systems Act 32 of 2000. Gazette No. 21776, Notice No. 1187, Government Printers, Pretoria (20 November)

RSA (Republic of South Africa). 2004. National Environmental Management: Air Quality Act 39 of 2004. Government Gazette vol. 476 no. 27318. Government Printers, Pretoria (25 February 2005).

RSA (Republic of South Africa). 2008. National Environmental Management: Waste Act 59 of 2008. Government Gazette, vol. 525, no. 32000. Government Printers, Pretoria (10 March).

Smith, JE. 1994. Biotechnology handbooks: Asperigillus. Springer Science and Business Media LLC.

Snyman, J. 2009. A zero waste model for the City of Tshwane Metropolitan Municipality. Doctor Technologiae. Pretoria: Tshwane University of Technology.

South Africa: Consolidated Regulations. 2003. Occupational Health and Safety Act 85 of 1993 – Regulations and Notices – Government Notice R2281. Available at: http://www.saflii.org/za/legis/consol_reg/ohasa85o1993rangnr2281716 [Accessed 29 June 2016].

Taiwo, AD. 2011. Composting as a Sustainable Waste Management Technique in Developing Countries. *Journal of Environmental Science and Technology*, 4, 93–102.

Williams, PT. 2005. Waste treatment and disposal (2nd ed.). West Sussex [England]: John Wiley.

Zurbrügg, C, Drescher, S, Patel., A and Sharatchandra, HC. 2004. Decentralised composting of urban waste: an overview of community and private initiatives in Indian cities. *Waste Management*, 24(7), 655–662.

A BIOREFINERY APPROACH TO IMPROVE THE SUSTAINABILITY OF THE SOUTH AFRICAN SUGAR INDUSTRY: AN ASSESSMENT OF SELECTED SCENARIOS

K Haigh, MA Mandegari, S Farzad, AG Dafal and JF Görgens

Department of Process Engineering, Stellenbosch University,

Stellenbosch, 7602, South Africa.

Corresponding author e-mail: khaigh@sun.ac.za

ABSTRACT

The potential for implementation of biorefineries, annexed to South African sugar mills was investigated to determine if this would improve sustainability of the sugar industry. There are opportunities to make more efficient use of the sugar cane plant as the bagasse is currently burnt in inefficient boilers and the leaves are burnt during harvesting. Simulations of selected biorefinery scenarios were generated using Aspen Plus®. The selected scenarios included a cellulosic ethanol process as the baseline. The potential to improve the economics of the ethanol scenario by co-production of higher value furfural and lactic acid were considered in two further scenarios. Furthermore, scenarios covering methanol synthesis, Fischer-Tropsch (FT) synthesis and butanol synthesis have been developed. In addition to chemicals, electricity for export to the grid was a co-product for all scenarios. The data from these simulations was used to generate an economic assessment of each scenario and subsequently used to carry out a sustainability assessment in terms of economic, environmental and social parameters. From the scenarios investigated, it was determined that biorefineries have the potential to reduce environmental impacts and create jobs. However, the economics of the investigated scenarios are not sufficiently robust to justify investment. There are multitudes of other scenarios that require consideration and there is ongoing technological development in this area.

Keywords: sugar mill biorefinery, ethanol, methanol, lactic acid, furfural, Fischer-Tropsch

INTRODUCTION

It has been proposed that the sustainability of the South African sugar industry can be improved by product diversification through sugar cane biorefineries using agricultural and processing residues. Approximately 1 million people in South Africa depend directly and indirectly on the current sugar industry for their livelihoods. This includes

25000 registered small-scale growers on the cultivation side (SASA). Biorefining is the sustainable processing of biomass into a spectrum of marketable products and energy (Cherubini, 2010a) and embraces a wide range of technologies able to separate biomass resources (wood, grasses, corn, sugar cane, etc.) into their building blocks (carbohydrates, proteins, triglycerides, etc.), which can be converted to value-added products, such as materials, biofuels, chemicals and electricity as part of a biobased economy. The idea of a biobased or green economy is that the focus shifts towards producing fuels, chemicals and advanced materials from biomass in a way that supports development of the existing rural community and existing biobased industries while minimising harm to the environment (Langeveld et al., 2010).

In addition to sucrose, which is used to make sugar, the sugar cane plant consists of a large amount of fibrous material, which includes parts of the stalks and the leaves. This fibrous material is typically referred to as lignocellulosic biomass and is typically composed of cellulose (40-45%), hemicelluloses (20-30%), lignin (10-25%), ash and minor components (extractives, acids, salts and minerals (Balat, 2011; Benjamin et al., 2013). A number of opportunities have been identified within the sugar industry to access lignocellulosic biomass. Current sugar cane harvesting practices involve burning the majority of sugar cane leaves and tops, emitting high levels of particulate matter, CO, NO_x, SO_x and CH_4 (Leal et al., 2013), that cause serious health problems for workers who breathe in the soot while working, as well as those who live in nearby areas. These sugar cane residues, leaves and tops, could be harvested separately as a feedstock for production of value-added products. Bagasse is burnt to provide process energy for the sugar mills. However, most South African sugar mills use old and inefficient boilers to produce low pressure steam for process energy (Leibbrandt et al., 2011). Improving the efficiency of the existing boilers will reduce the amount of sugar cane bagasse required for process energy and provide additional lignocellulosic material for conversion to chemicals and excess electricity (for export sales from the sugar mill).

For lignocellulosic biomass, the technology for the production of ethanol is commercially better established than technologies for other biofuels and biochemicals (Demirbas, 2009; Devarapalli & Atiyeh, 2015; Fatih Demirbas, 2009; Karlsson et al., 2014). The main factor inhibiting the production of ethanol from lignocellulosic biomass is that this fuel is expensive when compared to conventional gasoline. The lignocellulosic biomass that can be used to produce ethanol is recalcitrant and requires harsh pre-treatment conditions in order to ensure that the cellulose is accessible for the subsequent hydrolysis to the fermentable sugars required for ethanol production (Balat, 2011; Haghighi Mood et al., 2013).

Various options to broaden the product range and improve the economic viability of ethanol production from lignocellulose have thus been investigated to make more efficient use of the biomass (Jingura & Kamusoko, 2015). Biological conversions can be used to produce a variety of products including lactic acid, succinic acid and itaconic acid (Menon & Rao, 2012). South Africa is currently one of the world's largest producers of an existing biobased product, furfural (De Jong & Marcotullio, 2010). This is an example of a chemical conversion process using an acid catalyst (Zeitsch, 2000).

Thermochemical conversion of lignocellulose to biofuels via gasification-synthesis is an alternative to biological conversion (Consonni *et al.*, 2009). Gasification-synthesis converts both carbohydrates and lignin to biofuel. Intermediate liquid fuel products such as biomethanol or biosyncrude can take advantage of existing fuel production infrastructure, through processing with fossil-based fuels at existing refineries, producing Fischer-Tropsch (FT) liquids or biodiesel. These two gasification-based scenarios have been proposed to evaluate thermochemical conversion as an alternative to biological conversion routes. Gasification will be used to produce syngas, which will subsequently be converted using FT synthesis or methanol synthesis, based on well-established processes for the conversion of natural gas.

A schematic representation of the options available to produce chemicals from biomass is shown in Figure 4.1. A detailed investigation is required to develop a better understanding of which chemicals and processes are most suitable for the development of biorefineries based on sugar cane residues. The scenarios were developed on the basis that the new biorefineries would be annexed to an existing sugar mill. In each case, the existing boiler will be replaced with a new, efficient energy island (combined heat and power plant (CHP) to supply process energy to both the sugar mill and the new biorefinery. The bagasse will be mixed with the harvest residues and diverted to the new energy island and biorefinery. It has been assumed that a portion of the harvest residue will be available due to a change in harvesting practices, which means that burning of sugar cane prior to harvesting will cease. The aim of this work is to carry out a comprehensive assessment of the potential of biorefineries in South Africa and thus, six scenarios have been identified as examples of potential biorefineries. A process simulation of each scenario will be developed using Aspen Plus® (Ali Mandegari *et al.*, 2016). This data will be used to carry out a sustainability assessment, which is comprised of economic, environmental and social indicators of the investigated scenarios.

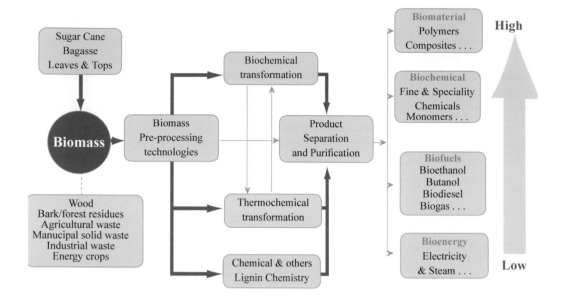

Figure 4.1:   Schematic representation of sugar cane biorefinery options

MATERIALS AND METHODS

Feedstock composition

The biomass availability was calculated in terms of the agricultural residues that would be available for such a biorefinery, assuming that only the "brown leaves" of the sugar cane plant will be utilised, while the green tops will be left in the field for nutrient retention in the soil. The calculation of the total lignocellulose and harvesting residues available is presented in Table 4.1. The composition of this type of feedstock has been reported and this information was used in this research as well (Benjamin, 2014; Benjamin *et al*., 2013; Smithers, 2014).

Table 4.1: Bagasse and harvesting residues production of a typical sugar cane mill

Material	Portion of material (%)	Flow rate (t/h)
Sugar cane		300
Wet bagasse	30% of sugar cane	90
Dry bagasse*	50% of wet bagasse	45
Brown leaves and green tops (total harvesting residues)	15% of sugar cane	45
Brown leaf available to biorefinery	50% of harvesting residues	22.5
Dry harvest residues	15% of wet	20
Total dry feedstock*		65

Extractives are included in the dry basis

Scenarios selected for investigation

In order to investigate the potential of sugar cane biorefineries, six scenarios were selected as examples for detailed investigation. They are summarised in Table 4.2. A simplified flow diagram of each scenario is given in Figure 4.2. For each flow diagram, the biomass is a mixture of sugar cane bagasse from the sugar mill and harvest residues. It is expected that a portion of the steam and electricity will be returned to the sugar mill. Process simulations were developed using Aspen Plus® flow sheeting software data available in literature.

Table 4.2: Summary of the selected scenarios

Scenario number	Main products	Energy products
Scenario 1	Bioethanol	Electricity
Scenario 2	Bioethanol and lactic acid	Electricity
Scenario 3	Bioethanol and furfural	Electricity
Scenario 4	Biomethanol	Electricity
Scenario 5	Fischer-Tropsch biofuel	Electricity
Scenario 6	Biobutanol	Electricity

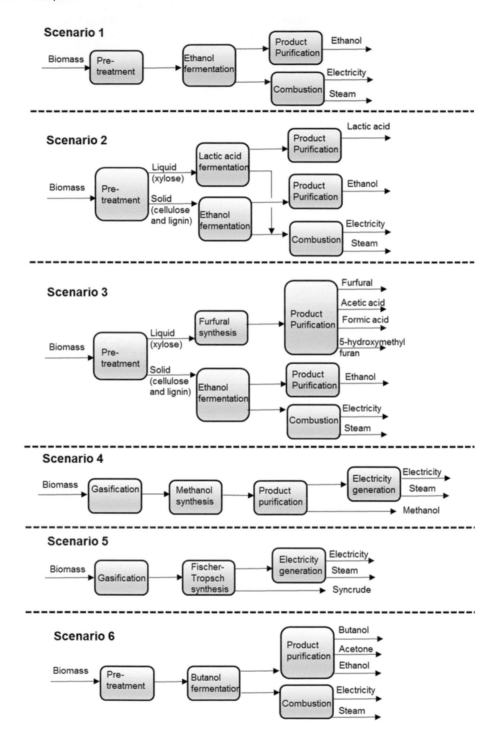

Figure 4.2:   Schematic representation of the investigated biorefinery scenarios

Description of the ethanol baseline

The minimum essential process steps for bioethanol production from lignocellulose via the biochemical route are pre-treatment, conditioning (in some cases), hydrolysis and fermentation, followed by conventional steps in purification (typically distillation and dehydration). Prior to the enzymatic hydrolysis process, it is necessary to pre-treat the material to make the cellulose accessible for the enzymes. The pre-treatment techniques can be divided into four different categories, mechanical, mechanical-chemical, chemical and biological (Alvira *et al.*, 2010; Haghighi Mood *et al.*, 2013). Techniques such as steam explosion and dilute acid pre-treatment are considered to be near readiness for industrial application, being similar to processes already used in the paper and pulp industry. For this reason, steam explosion with sulphur dioxide impregnation was investigated (Carrasco *et al.*, 2010). Following pre-treatment, the next step is hydrolysis, which means the cleaving of a cellulose molecule by adding a water molecule. This reaction (cellulose to glucose) can be catalysed by dilute acid, concentrated acid or enzymes, although enzymes are preferred (Gupta & Verma, 2015; Rabelo *et al.*, 2011).

The fermentation process selected for ethanol fermentation was simultaneous saccharification and co-fermentation (SSCF), which means that a single vessel is used for the fermentation of both glucose and xylose thus reducing the number of reaction vessels required and avoiding product inhibition issues associated with enzymes (Balat *et al.*, 2008; Hamelinck *et al.*, 2005; Mosier *et al.*, 2005; Olofsson *et al.*, 2008). Thus, for the ethanol baseline, the whole slurry for the pre-treatment process was fermented to produce ethanol.

Options to improve ethanol yields by adding molasses to the fermentation broth are currently being investigated (Dias *et al.*, 2012; Gnansounou *et al.*, 2015). However, the focus of this project was to consider the use of lignocellulosic biomass for a range of valorisation options and molasses would not be a suitable feedstock for all of the investigated scenarios. Thus, this approach has been deemed outside the scope of this project.

The purification and recovery section separates water, anhydrous ethanol and combustible solids from the fermentation broth. Ethanol distillation is a widely investigated topic due to its high impact on the total energy consumption of the plant (Batista *et al.*, 2012; Dias *et al.*, 2011; Errico *et al.*, 2013). Although, many attempts have been made to develop an efficient bioethanol purification process, in all studies the main equipment configuration is almost the same (Batista *et al.*, 2012).

In order to establish a comprehensive biorefinery, in addition to the bioethanol production section, other supplementary units such as evaporation, wastewater treatment, boiler and steam/power generation are required. Stillage water is purified for reuse in the process, using an evaporation unit. Since the amount of stillage water that can be recycled is limited, an evaporation unit would be more effective than a wastewater treatment unit. In this context, a multiple effect evaporator is used to treat the stillage water rather than purifying the stillage water in the wastewater treatment unit. The ethanol process generates a number of wastewater streams that must be treated before being recycled to the process or released to the environment. This is accomplished in the wastewater treatment (WWT) unit. Boiler and power generation units are described later in this chapter.

Description of the ethanol-lactic acid scenario

The conventional processes for producing lactic acid from lignocellulosic biomass includes four main steps (Taherzadeh & Karimi, 2008), namely; pre-treatment, enzymatic hydrolysis, fermentation, separation and purification of lactic acid to meet the standards of commercial applications.

In this study, steam explosion pre-treatment is utilised as it has been investigated extensively in recent years. From the integrated second-generation biorefinery perspective, co-production of lactic acid with ethanol is considered, where ethanol is produced from the cellulose while lactic acid is produced from the hemicellulose fraction.

During the fermentation process, the produced lactic acid is continuously neutralised using magnesium hydroxide [$Mg(OH)_2$] to minimise its inhibitory effect on the fermenting strains forming Mg-lactate. The latter will react with a water-miscible organic amine, triethylamine(R_3N), forming $Mg(OH)_2$ crystals and a triethylamine-Lactate complex, (R_3N-Lactate) in an exchange reactor referred to as a SWAP reactor. The $Mg(OH)_2$ crystal will be filtered and recycled back to fermentation and the triethylamine-lactate complex will be thermally decomposed to release lactic acid where triethylamine will be recycled back to the SWAP reactor.

In the present work, ethanol is used as it is available in an integrated biorefinery from lignocellulosic biomass. Lactic acid is first reacted with ethanol to form a more volatile ethyl lactate, which can be more easily purified by distillation as the top product, while the heavy organic acids and impurities are collected as the bottom product. The ethyl lactate is then hydrolysed in a second reactive distillation column where lactic acid is collected at the bottom, while ethanol and water is produced at the top. Ethanol will be separated from water by normal distillation and recycled back to the esterification unit. Although the ethanol and lactic acid production processes are different, the evaporation, WWT, boiler and power generation units are integrated to handle streams of both processes.

Description of the ethanol-furfural scenario

Current commercial scale processes used the whole biomass as the feedstock for furfural production by means of an acid catalysed reaction although only the pentose sugars in the hemicellulose fraction are converted to furfural (Xing et al., 2011; Zeitsch, 2000). However, this degrades the cellulose fraction which could be used to produce ethanol (De Jong & Marcotullio, 2010). In this scenario, the baseline scenario was modified so that the liquid, hemicellulose fraction was separated from the solid cellulose fraction and used to make furfural. The solid fraction was used to make ethanol. The furfural process uses tetrahydrofuran as an extractive solvent to remove the furfural from the reaction mixture quickly as this should prevent side reactions and improve the yield (Xing et al., 2011). After the reaction was completed, a decanter was used to separate the aqueous and organic phases. A series of distillation columns was used to separate and purify the final products, which included furfural, acetic acid, formic acid and 5-hydroxymethyl furan. The tetrahdrofuran was recovered and recycled.

Description of the biomethanol scenario

In advanced methanol synthesis investigated in this study, conditioned synthesis gas from the Rectisol™ unit enters the methanol synthesis section where it is compressed to 98 bar (liquid phase) in the synthesis reactor (Hamelinck & Faaij, 2002). The reactor feed is pre-heated to the reaction temperature of 250°C, which is isothermally maintained by steam generation. Furthermore, cooling of the reactor effluent is achieved by heating up the reactor feed directly. The vapour-phase product from the synthesis reactor must be cooled to recover the methanol and to allow unconverted syngas and any inert gaseous species (CO_2, CH_4) to be separated (Diederichs et al., 2016). Cooling water is used to lower the temperature to 32°C, the temperature at which the majority of the liquid methanol condenses and is separated in a knock-out vessel.

The methanol is still at elevated pressure at this point in the process, resulting in a significant quantity of gas being absorbed in the methanol stream as it leaves the synthesis section of the process. These gases are removed from the methanol at this stage of the process. The combined gas streams are heated before expansion through a turbo expander generator to recover some of the compression energy of the gas by generating electricity.

Description of the FT scenario

As an alternative to converting the conditioned syngas from biomass gasification into methanol, the syngas can also be converted to a range of high quality fuels and chemicals using FT synthesis.

The advanced FT reactor working in a low temperature regime at 40 bar and 240°C, is simulated in this research, while its once-through conversion per pass is specified as 80% (Kreutz et al., 2008). Regarding the product distribution of the hydrocarbons formed, an equal distribution between the four product types (i.e. gases, naphtha, diesel and waxes) is specified, which occurs at the assumed probability of 90% on the Anderson-Schulz-Flory distribution (Ekbom et al., 2009).

Description of the butanol scenario

Butanol can be produced from biomass by means of an acetone-butanol-ethanol (ABE) fermentation using strains of *Clostridia* in an oxygen free environment (Dürre, 2008; García et al., 2011; Karimi & Pandey, 2014). Challenges with this process include feed and product inhibition, thus SSF fermentation combined with gas stripping was identified as the most suitable approach to achieve high yields (Qureshi et al., 2010; Qureshi et al., 2014). The solvents can be recovered from the gas by means of condensation and then sent for further purification (Ezeji et al., 2005; Qureshi et al., 2014). Liquid-liquid extraction was used to extract butanol, ethanol and acetone from the condensate. This concentrated solvent mixture was then sent to the distillation train for product recovery and purification.

Provision of process energy

The energy demand of the scenarios has been evaluated based on the combined utility requirements of the processes of the sugar mill and the annexed biorefinery. The sugar industry infrastructure is old and different technologies with different energy demand are available. However, an efficient sugar mill with steam consumption of 40 ton/100 ton of cane is considered in this work.

In some cases, the amount of organic by-product available from the biorefinery, fed to the energy island, cannot produce sufficient steam and electricity to meet the combined demands of the biorefinery and sugar mill. Combustion of a portion of the bagasse and harvest residue feedstocks in the boiler is the solution favouring the environmental benefits of the biorefinery. This will provide sufficient process energy for the sugar mill and biorefinery, but causes a corresponding decrease in the amount of feedstock available for biorefinery products, which has negative economic impacts (smaller production scales are negatively affected by economies of scale; less revenue generated from the biorefinery per ton of harvested cane).

Economic assessment and the key economic parameters

An economic assessment of the developed biorefineries is conducted by estimation of purchased and installed cost of equipment, variable operating cost, fixed operating cost, economic parameters and selection of representative parameters – all of which are based on technical information obtained from process simulations.

In general, the price of equipment can be defined from the Aspen Plus® Economic Evaluator package or using real quotes based on published literature and reports. In this research, equipment costs are mostly estimated using the Aspen Plus® Economic Evaluator package (Aspen Technology, Inc., USA) especially flash drums, columns, pumps, compressors, and heat-exchangers. The results of Aspen Plus® for special equipment such as boilers, turbo-expanders, generators, reactors and wastewater treatment basins are not reliable, thus technical report data was used (Aiden et al., 2002; Humbird et al., 2011).

The operating cost is divided into two categories as follows:

- **Variable operating cost** is the cost of feedstock, chemicals and disposal waste. This varies according to plant capacity. Variable operating cost is calculated according to the mass flow rate of the streams and market prices.
- **Fixed operating cost** is the fixed cost of the plant regardless of the plant operational capacity, after design. The main sources of the fixed operating cost are employee salaries, maintenance (3% of inside battery limits), and property insurance and tax (0.7% of fixed capital investment).

Sustainability assessment

Sustainability is typically regarded as having three components or pillars, namely: economic viability, environmental impact, and social benefits (Martins et al., 2007).

The system boundary applied to this work was from "cradle to gate", encompassing the cultivation and harvesting of sugar cane, the sugar mill, the operation of the biorefinery and the combined heat and power plant (also referred to as the energy island). The research methodology applied in this study was based on Standard ISO14040 (ISO, 2006a) and ISO14044 (ISO, 2006b), which sets out a standardised approach to assess the environmental impacts of the biorefineries.

The identification of relevant sustainability issues associated with integrated biorefineries is crucial for the development of sustainability indicators and subsequent assessment of these facilities and their products needs to take into account the expectations of the stakeholders. A life cycle approach covering the whole supply chain was used. In this study –

- The *environmental indicators* used are those environmental impact indicators of life cycle assessment (LCA). LCA will be used as a tool to assess the environmental sustainability of biorefineries (Cherubini, 2010b).

- The *economic sustainability assessment* used the NPV as the key parameter from the economic assessment. This is essentially an investment analysis to indicate whether the biorefinery will yield sufficient economic returns to justify investment of private capital, which is essential for development of the green economy.

- The *social indicators* considered included employment opportunities, health and safety, and local community impacts.

RESULTS AND DISCUSSION

Six potential biorefineries were identified as examples to investigate the potential to broaden the range of products from the sugar industry and thus improve sustainability. Overall, the data indicates that there is significant potential to improve the sustainability of the sugar mills, which includes substantial environmental benefits when compared to the fossil-derived products. Given that sustainability consists of three sets of indicators, the data has been normalised for comparison. The resulting plot is given in Figure 4.3.

It was found that the greatest environmental benefits were derived from the combined ethanol and lactic acid scenario; this was followed by methanol synthesis, butanol production, FT synthesis, cellulosic ethanol, while the lowest environmental benefits were derived from the combined ethanol furfural scenario.

The results show that there is a potential for job creation with the bulk of the jobs in the supply of feedstocks as harvesting revenues. It is expected that the amount of jobs created in the supply of feedstocks will remain the same for each scenario; however, the number of jobs created in the biorefinery will vary. It was found that the largest number of biorefinery jobs would be created in the combined ethanol and lactic acid and the combined ethanol and furfural scenarios, which can be attributed to the additional complexity of these processes. The smallest number of jobs would be created in the FT synthesis scenario.

In order to be deemed economically viable, the scenarios need to meet minimum investment criteria. This work found that three of these scenarios had an NPV above the minimum threshold. This was calculated to be 0.61 on a normalised basis. The methanol synthesis scenario had the highest NPV, followed by the combined ethanol and lactic scenario, and then the FT scenario; however, the economics of the FT scenario are highly dependent on the price of crude oil. Of the investigated scenarios, the ethanol along with electricity is the most technologically mature option but it falls short of the investment criteria.

For all the scenarios, it was observed that there was a trade-off between the three pillars of sustainability. Specifically, the methanol synthesis scenario has the highest NPV along with an average environmental performance and a low job creation potential, while the combined lactic acid ethanol scenario has a lower NPV but its environmental and job creation values are higher.

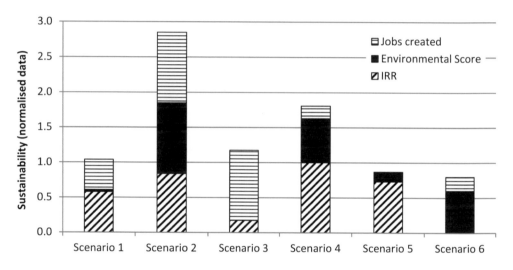

Figure 4.3:   A comparison of the sustainability of the investigated scenarios, based on normalised data

CONCLUSIONS

Biorefineries using residues from industrial biomass processing can provide substantial environmental benefits relative to the fossil-derived products that they replace, which justifies their inclusion in the green economy.

Biorefineries can provide significant job creation, mostly in the supply of feedstocks, such as the harvesting residues to the sugar mills. The so-called "green cane harvesting," as an alternative to burning of cane before harvesting, will essentially double the number of jobs in harvesting, while also creating additional jobs in collection and transport of these residues to sugar mills. There is a potential economic cost to green cane harvesting, due to additional jobs required, which must be mitigated through attractive economic returns from processing of these residues

The economic viability of biorefineries is critical to attracting private investment, while also securing environmental and social benefits. High value products and attractive manufacturing economics are required to secure the investment returns.

There are clearly conflicts between the three pillars of sustainability, for example, the improved investor returns of the methanol biorefinery, compared to the ethanol-lactic acid scenario, come at the expense of reduced environmental and social benefits – such conflicts and compromises are typical in sustainable development.

ACKNOWLEDGEMENTS

We would like to thank the following institutions for their support: Aspen Technology Inc. for provision of the Aspen Plus® academic licences used to carry out this work. Aspen Plus® is a registered trademark of Aspen Technology Inc., the National Research foundation for providing K. Haigh with a bursary; and SASRI for providing data on sugar cane cultivation. This chapter forms part of a research project, "Greener Cities in South Africa", funded by the Green Fund, an environmental finance mechanism implemented by the Development Bank of Southern Africa (DBSA) on behalf of the Department of Environmental Affairs (DEA). Opinions expressed and conclusions arrived at, are those of the author and are not necessarily to be attributed to the Green Fund, DBSA or DEA.

REFERENCES

Aiden, A, Ruth, M, Ibsen, K, Jechura, J, Neeves, K, Sheehan, J and Wallace, B. 2002. Lignocellulosic Biomass to Ethanol Process Design and Economics Utilizing Co-Current Dilute Acid Prehydrolysis and Enzymatic Hydrolysis for Corn Stover (NREL).

Ali Mandegari, M, Farzad, S and F Görgens, J. 2016. Process Design, Flowsheeting and Simulation of Bioethanol Production from Lignocellulose. In Biofuels: Production and Future Perspectives, (CRC Press), pp. 261–287.

Alvira, P, Tomás-Pejó, E, Ballesteros, M and Negro, MJ. 2010. Pretreatment technologies for an efficient bioethanol production process based on enzymatic hydrolysis: A review. *Bioresource Technology*, 101, 4851–4861.

Balat, M. 2011. Production of bioethanol from lignocellulosic materials via the biochemical pathway: A review. *Energy Conversion and Management*, 52, 858–875.

Balat, M, Balat, H and Öz, C. 2008. Progress in bioethanol processing. *Progress in Energy and Combustion Science*, 34, 551–573.

Batista, FRM, Follegatti-Romero, LA, Bessa, LCBA, and Meirelles, AJA. 2012. Computational simulation applied to the investigation of industrial plants for bioethanol distillation. *Computers & Chemical Engineering*, 46, 1–16.

Benjamin, YL. 2014. Sugarcane cultivar selection for ethanol production using dilute acid pretreatment, enzymatic hydrolysis and fermentation.

Benjamin, Y, Cheng, H and Görgens, JF. 2013. Evaluation of bagasse from different varieties of sugarcane by dilute acid pretreatment and enzymatic hydrolysis. *Industrial Crops and Products*, 51, 7–18.

Carrasco, C, Baudel, HM, Sendelius, J, Modig, T, Roslander, C, Galbe, M, Hahn-Hägerdal, B, Zacchi, G.and Lidén, G. 2010. SO2-catalyzed steam pretreatment and fermentation of enzymatically hydrolyzed sugarcane bagasse. *Enzyme and Microbial Technology*, 46, 64–73.

Cherubini, F. 2010a. The biorefinery concept: Using biomass instead of oil for producing energy and chemicals. *Energy Conversion and Management*, 51, 1412–1421.

Cherubini, F. 2010b. The biorefinery concept: Using biomass instead of oil for producing energy and chemicals. *Energy Conversion and Management*, 51, 1412–1421.

Consonni, S, Katofsky, RE and Larson, ED. 2009. A gasification-based biorefinery for the pulp and paper industry. *Chemical Engineering Research and Design*, 87, 1293–1317.

De Jong, W and Marcotullio, G. 2010. Overview of biorefineries based on co-production of furfural, existing concepts and novel developments. *International Journal of Chemical Reactor Engineering*, 8.

Demirbas, A. 2009. Biorefineries: Current activities and future developments. *Energy Conversion and Management*, 50, 2782–2801.

Devarapalli, M and Atiyeh, HK. 2015. A review of conversion processes for bioethanol production with a focus on syngas fermentation. *Biofuel Research Journal*, 2, 268–280.

Dias, MOS, Junqueira, TL, Jesus, CDF, Rossell, CEV, Maciel Filho, R and Bonomi, A. 2012. Improving second generation ethanol production through optimization of first generation production process from sugarcane. *Energy*, 43, 246–252.

Dias, MOS, Modesto, M, Ensinas, AV, Nebra, SA, Filho, RM and Rossell, CEV. 2011. Improving bioethanol production from sugarcane: evaluation of distillation, thermal integration and cogeneration systems. *Energy*, 36, 3691–3703.

Diederichs, GW, Mandegari, MA, Farzad, S and Görgens, JF. 2016. Techno-economic comparison of biojet fuel production from lignocellulose, vegetable oil and sugar cane juice. Bioresource Technology.

Dürre, P. 2008. Fermentative Butanol Production: Bulk Chemical and Biofuel. *Annals of the New York Academy of Sciences*, 1125, 353–362.

Ekbom, T, Hjerpe, C, Hagström, M and Hermann, F. 2009. Pilot study of Bio-jet A-1 fuel production for Stockholm-Arlanda Airport. Stockholm: VÄRMEFORSK Service AB.

Errico, M, Rong, B.-G, Tola, G and Spano, M. 2013. Optimal synthesis of distillation systems for bioethanol separation. Part 2. Extractive distillation with complex columns. *Industrial & Engineering Chemistry Research*, 52, 1620–1626.

Ezeji, TC, Karcher, PM, Qureshi, N and Blaschek, HP. 2005. Improving performance of a gas stripping-based recovery system to remove butanol from Clostridium beijerinckii fermentation. *Bioprocess and Biosystems Engineering*, 27, 207–214.

Fatih Demirbas, M. 2009. Biorefineries for biofuel upgrading: A critical review. *Applied Energy*, 86, Supplement 1, S151–S161.

García, V, Päkkilä, J, Ojamo, H, Muurinen, E and Keiski, RL. 2011. Challenges in biobutanol production: How to improve the efficiency? *Renewable and Sustainable Energy Reviews*, 15, 964–980.

Gnansounou, E, Vaskan, P and Pachón, ER. 2015. Comparative techno-economic assessment and LCA of selected integrated sugarcane-based biorefineries. *Bioresource Technology*, 196, 364–375.

Gupta, A and Verma, JP. 2015. Sustainable bio-ethanol production from agro-residues: A review. *Renewable and Sustainable Energy Reviews*, 41, 550–567.

Haghighi Mood, S, Hossein Golfeshan, A, Tabatabaei, M, Salehi Jouzani, G, Najafi, GH, Gholami, M and Ardjmand, M. 2013. Lignocellulosic biomass to bioethanol, a comprehensive review with a focus on pretreatment. *Renewable and Sustainable Energy Reviews*, 27, 77–93.

Hamelinck, CN and Faaij, AP. 2002. Future prospects for production of methanol and hydrogen from biomass. *Journal of Power Sources*, 111, 1–22.

Hamelinck, CN, Hooijdonk, G and Van Faaij, APC. 2005. Ethanol from lignocellulosic biomass: techno-economic performance in short-, middle- and long-term. *Biomass and Bioenergy*, 28, 384–410.

Humbird, D, Davis, R, Tao, L, Kinchin, C, Hsu, D and Aden, A. 2011. Process design and economics for biochemical conversion of lignocellulosic biomass to ethanol.

ISO 2006a. ISO 14040 – Environmental management life cycle assessment – Principles and framework.

ISO 2006b. ISO 14044 – Environmental management life cycle assessment – Requirements and guidelines.

Jingura, RM and Kamusoko, R. 2015. A multi-factor evaluation of Jatropha as a feedstock for biofuels: the case of sub-Saharan Africa. *Biofuel Research Journal*, 2, 254–257.

Karimi, K and Pandey, A. 2014. Current and future ABE processes. *Biofuel Research Journal*, 3,77.

Karlsson, H, Börjesson, P, Hansson, P.-A and Ahlgren, S. 2014. Ethanol production in biorefineries using lignocellulosic feedstock – GHG performance, energy balance and implications of life cycle calculation methodology. *Journal of Cleaner Production*, 83, 420–427.

Kreutz, TG, Larson, ED, Liu, G and Williams, RH. 2008. Fischer-Tropsch fuels from coal and biomass. In 25th Annual International Pittsburgh Coal Conference.

Langeveld, JWA, Dixon, J and Jaworski, JF. 2010. Development perspectives of the biobased economy: A Review. *Crop Science*, 50, S-142-S-151.

Leal, MRLV, Galdos, MV, Scarpare, FV, Seabra, JEA, Walter, A and Oliveira, COF. 2013. Sugarcane straw availability, quality, recovery and energy use: A literature review. *Biomass and Bioenergy*, 53, 11–19.

Leibbrandt, NH, Knoetze, JH and Görgens, JF. 2011. Comparing biological and thermochemical processing of sugarcane bagasse: An energy balance perspective. *Biomass and Bioenergy*, 35, 2117–2126.

Martins, AA, Mata, TM, Costa, CAV and Sikdar, SK. 2007. Framework for Sustainability Metrics. *Industrial & Engineering Chemistry Research*, 46, 2962–2973.

Menon, V, and Rao, M. 2012. Trends in bioconversion of lignocellulose: Biofuels, platform chemicals & biorefinery concept. *Progress in Energy and Combustion Science*, 38, 522–550.

Mosier, N, Wyman, C, Dale, B, Elander, R, Lee, YY, Holtzapple, M and Ladisch, M. 2005. Features of promising technologies for pretreatment of lignocellulosic biomass. *Bioresource Technology*, 96, 673–686.

Olofsson, K, Bertilsson, M, Lidén, G and others. 2008. A short review on SSF-an interesting process option for ethanol production from lignocellulosic feedstocks. *Biotechnol Biofuels*, 1, 1–14.

Qureshi, N, Cotta, MA and Saha, BC. 2014. Bioconversion of barley straw and corn stover to butanol (a biofuel) in integrated fermentation and simultaneous product recovery bioreactors. *Food and Bioproducts Processing*, 92, 298–308.

Qureshi, N, Saha, BC, Dien, B, Hector, RE and Cotta, MA. 2010. Production of butanol (a biofuel) from agricultural residues: Part I – Use of barley straw hydrolysate. *Biomass and Bioenergy*, 34, 559–565.

Rabelo, SC, Amezquita Fonseca, NA, Andrade, RR, Maciel Filho, R and Costa, AC. 2011. Ethanol production from enzymatic hydrolysis of sugarcane bagasse pretreated with lime and alkaline hydrogen peroxide. *Biomass and Bioenergy*, 35, 2600–2607.

SASA The South African Sugar Industry Directory.

Smithers, J. 2014. Review of sugarcane trash recovery systems for energy cogeneration in South Africa. *Renewable and Sustainable Energy Reviews*, 32, 915–925.

Taherzadeh, MJ and Karimi, K. 2008. Pretreatment of lignocellulosic wastes to improve ethanol and biogas production: A review. *International Journal of Molecular Sciences*, 9, 1621–1651.

Xing, R, Qi, W and Huber, GW. 2011. Production of furfural and carboxylic acids from waste aqueous hemicellulose solutions from the pulp and paper and cellulosic ethanol industries. *Energy & Environmental Science*, 4, 2193.

Zeitsch, KJ. 2000. The chemistry and technology of furfural and its many by-products. Amsterdam; New York: Elsevier.

INTEGRATED WASTE MANAGEMENT APPROACH: USE OF ACTI-ZYME FOR MUNICIPAL SEWAGE TREATMENT AND RECOVERY OF BIOGAS AND BIOSOLIDS

MM Manyuchi,[1,2] **DIO Ikhu-Omoregbe**[1] **and OO Oyekola**[1]

[1] Department of Chemical Engineering, Cape Peninsula University of Technology, Cape Town, 7530, South Africa

[2] Department of Chemical and Process Systems Engineering, Harare Institute of Technology, Harare, Zimbabwe

Corresponding author e-mail: mmanyuchi@hit.ac.zw

ABSTRACT

Water and energy scarcity are global challenges, hence the need for sustainable wastewater management. Municipal sewage (MS) is often disposed of into river bodies, untreated. Additionally, the conventional disposal of MS sludge (MSS), a by-product from the MS treatment process, results in landfilling problems. This study focused on the anaerobic treatment of MS, co-harnessing biogas and biosolids as value-added products, utilising Acti-zyme, an enzyme biocatalyst. A techno-economic analysis was carried out to evaluate the viability of applying this technology on a large scale. An MS plant in Zimbabwe, with a capacity of 19.6 ML/day, operating efficiency of 60% and a lifespan of 20 years was considered for techno-economic assessment. MS treatment using Acti-zyme resulted in >60% removal of contaminants. At optimal treatment conditions resulted in production of biogas containing 78% of biomethane, together with biosolids containing 8.0, 5.0 and 1.0% of NPK respectively. A net financial benefit of US$5.7 million per annum and net energy production of 1.4 MWh were forecasted. An investment of US$22 million will be required for starting the project. A positive net present value of US$1.2 million with an internal rate of return of 17.6% and a payback period of 5.9 years were estimated. For breakeven to be realised, 183 MWh must be produced. The techno-economic assessment indicated the viability of treating MS using Acti-zyme co-harnessing biogas and biosolids as valued added products.

Keywords: Acti-zyme, biogas, biosolids, integrated municipal sewage treatment, techno-economic analysis

INTRODUCTION

Water scarcity and pollution are global issues. The majority of those affected live in developing countries. A good example of a municipality facing wastewater treatment challenges is Chitungwiza; a satellite town in Zimbabwe, where municipal sewage is disposed of without adequate treatment, into the receiving water body, which is one of the major rivers providing drinking water to the residents of Chitungwiza and Harare, Zimbabwe (Nhapi, 2009). The disposal of untreated municipal wastewater poses health risks and environmental hazards (Nhapi, 2009). Furthermore, there are challenges with the municipal sewage sludge (MSS) management, which is generated during the MS treatment (MST). Therefore, there is a need to find an economic and alternative method for treating this MS. Employing biological means, such as anaerobic digestion mediated by Acti-zyme, which are environmentally friendly, sustainable and have a potential to generate value-added products meet these requirements. Acti-zyme is a biocatalyst consisting of a combination of bacterium and enzymes that has been successfully used for treating wastewater in numerous parts of the world including Australia, Canada and Zimbabwe (Tshuma, 2010). Acti-zyme was found to be an immotile biocatalyst. In addition, it contained several enzymes that can be used in sewage treatment, such as catalase, which has a detoxifying effect, protease, which breaks down proteins in sewage into amino acids, and amylase, which breaks down the complex sugars in sewage to simple sugars (Manyuchi *et al.*, 2015). Most waste management approaches employing Acti-zyme have been for environmental purposes in order to prevent pollution, preserve the ecosystem and protect public health (Cail, 1986; Dzvene, 2013). However, an integrated approach to the management of these wastes that focuses on energy recovery will have a significant economic and environmental impact. Biogas from wastes can be used as replacement or supplement for fossil fuels in power generation, while the by-products of biosolids are environmentally friendly fertilisers that can be used to enhance crop production. These are crucial for sustainable development in developing countries.

Acti-zyme is applicable under both aerobic and anaerobic conditions, reducing characteristic contaminants, such as total Kjeldahl nitrogen (TKN), total phosphates (TP), nitrates, ammonia, biological oxygen demand (BOD_5) and total suspended solids (TSS) by >40%. Furthermore, dissolved oxygen (DO) was shown to increase by >100% in treated dam water, promoting aquatic life (Tshuma, 2010). The use of Acti-zyme, as with any biological catalyst under anaerobic conditions, has the potential to favour biogas production (Manyuchi *et al.*, 2015). Although, Acti-zyme has been used for over five decades, its potential and suitability for biogas production in MS treatment has not received much attention

Although Acti-zyme has proven to be a successful biocatalyst for wastewater treatment, its application in municipal sewage treatment has not been well reported in open literature. Manyuchi *et al.*, 2015 showed that this biocatalyst exhibits optimal activity under mild conditions (37°C; pH 7). The optimal biogas and biosolid production conditions, and the techno-economic analysis to elucidate the viability of applying this technology in MST for the recovery of value-added products, have also not been reported. MST in many developing countries is inadequate. The accompanying MS sludge from many treatment technologies result in secondary pollution. Due to the water challenges faced by many developing country municipalities, there is a need to employ sustainable and integrated MST methods for water reuse, recycling and recovery of value-added products, which was the basis of this study.

MATERIALS AND METHODS

Application of Acti-zyme in MS treatment

Acti-zyme was obtained from Austech, the manufacturers of Acti-zyme, based in Australia, in September 2013. The MS was treated with Acti-zyme loadings of 0–0.060g/L for 0 and 70 days (Powell & Lundy, 2007). Temperature was fixed at 37°C and the agitation rate in the 500 mL flasks was maintained at 60 rpm. All parameters were measured using the American Public Health Association (APHA) standard methods of determination in wastewater (APHA, 2005). MS physico-chemical properties were measured for Acti-zyme free (A_o) MS effluent and for effluent at varying Acti-zyme loadings and retention times (A_1) to elucidate the impact of Acti-zyme on the MST. All experiments were conducted in replicates.

Raw MS sludge characterisation

The MS sludge (MSS) was filtered and dried to 60% to 80% moisture content. Moisture content and volatile matter analyses were done using an AND moisture analyser. The %moisture content (M) was determined by heating 5g of sample (105°C, 30 minutes) and then recording the difference in weight. The %volatile solids (VS) was determined by heating 5g of sample (105°C, 3 minutes), and then recording the difference in weight. MSS digestion was carried out at 37°C and 55°C to create mesophilic and thermophilic conditions respectively at atmospheric pressure.

Biogas quantity and quality measurement

The biogas quantity from the MSS digestion was measured in millilitres per day (mL/day) using the water displacement method. The biogas generated was taken from the sampling points for composition analysis. A GC 5400 gas chromatography analysis was used to analyse the biogas content and the composition was expressed as a percentage.

Biosolid quality measurement

The pH of the biosolids was measured using an HI 9810 Hanna pH electrode. The moisture content in the MSS was determined by heating 5g of MS sludge sample (105°C, 10 minutes) and then recording the difference in weight in an AND moisture analyser. After the digestion process, the biosolids were dried as a measure to reduce moisture content from 80% to 20%. The nitrogen, phosphorous and potassium (NPK) content in the biosolids was measured using a Labtronics double beam *uv-vis* spectrophotometer content. The *Eschericia coli* content was measured through the total plate count procedure (Lang *et al.*, 2007).

Techno-economic assessment approach

Capital budgeting techniques were used for the economic assessment. A local MS plant with a bionutrient removal plant capacity of 19.6 ML/day and an operating capacity of 60% was considered for potential MST using Acti-zyme, co-harnessing the biogas and the biosolids produced. A plant utilisation of 80% was considered, resulting in 292 operational days, being operated for 24 hours. Experimental data for the MS composition, biogas and biosolids generated were used as the basis for the techno-economic analysis. A total solids (TS) content of 1143 mg/L was considered based on the total amount of dissolved and suspended solids in the MS. In addition, a biogas production rate of 400 ml/day was considered with total biosolids generation amounting to about 10% of the TS in the MS, based on results from the current study.

Process description

Large objects were removed from the MS influent using a grit filter. The filtrate was fed to the primary settling tank whereby primary MSS is removed and fed to the anaerobic biodigester. Primary MSS does not comprise any Acti-zyme. The primary settling tank effluent is passed on to the secondary treatment stage (Figure 5.1). After the primary settling stage, Acti-zyme is added in the secondary clarifying tank at a loading rate of 0.050 g/L and a maximum retention time of 40 days and agitation is employed to increase Acti-zyme activity in the secondary settling tank. Settleable solids in the secondary clarifying tank are sent to the biodigester where they also act as feedstock for the biogas and biosolids production. These were termed secondary sludge as they contain Acti-zyme from the MST process (Figure 5.1). Traces of biogas produced from the anaerobic treatment of MS in the secondary clarifier are also sent to the biogas collection tank for purification. The MS effluent from the secondary settling tank is sent for chlorine disinfection to reduce pathogenic bacteria before final disposal to river systems. Alternatively, the MS effluent can be used for irrigation purposes.

The MSS material from the primary and secondary clarifying tanks is fed into a biodigester where Acti-zyme is added at 0.050 g/L. The retention time is 40 days. Biogas is produced for further separation into the different constituent gases. The biosolids formed during biogas production are dewatered and can be utilised as biofertilisers. On the other hand, the biogas produced is separated into constituent component with the biomethane being sent for electricity generation (Figure 5.1).

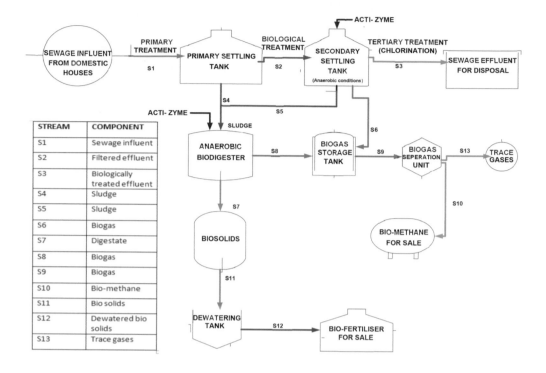

Figure 5.1: Process description for MS treatment using Acti-zyme co-harnessing biogas and biosolids

RESULTS AND DISCUSSION

Application of Acti-zyme in MS treatment

The raw MS physico-chemical characteristics obtained are indicated in Table 5.1. All the physico-chemical characteristics were significantly above the limits recommended by Environmental Management Agency (EMA) guidelines for domestic water and aquatic habitat (Nhapi, 2006). These characteristics show the potential toxicological effects of untreated MS disposal on the ecosystem. Decreases in the physico-chemical properties were achieved for Acti-zyme-free effluent (A_o), except for DO, which increased due to the removal of biocontaminants (Table 5.1). There was a clear indication that the effluent with Acti-zyme resulted in the physico-chemical properties that are more acceptable by EMA standards (Table 5.1). The decrease in the MS physico-chemical parameters in an Acti-zyme free system was attributed to the presence of native microbes and enzymes in the MS. A summary of the effect of optimal Acti-zyme loading (0.050 g/L) and a retention time (40 days) on MS physico-chemical parameters treated with Acti-zyme is given in Table 5.1. Biodegradation of municipal wastewater is often enhanced by the addition of specialised indigenous or commercial microbes (Yang *et al.*, 2014). This common practice is often targeted at pollution reduction without biogas recovery. The effect of Acti-zyme on MS treatment is shown in comparison to Acti-zyme-free MS. Table 5.2 compares the effect of Acti-zyme on different wastewaters. The results in Table 5.2 show that Acti-zyme has a potential to treat biologically contaminated wastewater due to its bioaugmentation characteristics whereby it enhances wastewater treatment in addition to the already existing natural micro-organisms.

Effect of Acti-zyme loading

Increase in Acti-zyme loading had a positive effect on the DO increase. This was attributed to the removal of all contaminants due to the Acti-zyme action. MS treated with Acti-zyme showed >60% reduction in the MS contaminants at optimum loadings of 0.050 g/L at a retention time of 40 days. The reduction in the parameter components is due to the uptake of nutrients in the MS by Acti-zyme, especially at increased loadings during its biocatalytic action. The DO concentration in the MS effluent increased by 91% to 87 mg/L with an increase in Acti-zyme loading for Acti-zyme loadings of 0-0.050 g/L, contrary to the Acti-zyme-free system (Table 5.1).

Table 5.1: MS characteristics for systems with Acti-zyme and without Acti-zyme with reference to disposal guidelines after 40 days (Values at optimal Acti-zyme loading of 0.05 g/L)

Parameter	Raw MS	MS effluent (A_o)	MS effluent (A_1)	EMA guidelines
TKN (mg/L)	245±5.5	39.9±0.20	9.4±0.36	10–20
BOD_5 @20°C (mg/L)	557±15.3	312.6±0.20	41.8±1.08	30–50
TSS (mg/L)	608±16.1	397.9±0.35	37.3±1.02	25–50
TDS (mg/L)	535±13	253.7±0.25	59.4±0.53	500–1500
EC @25°C (µS/cm)	3887±32.1	2186.2±0.21	1070.4±0.36	1000–2000
Cl- (mg/L)	833±11.2	673.6±0.50	263.3±4.02	-
pH @25°C	9±0.3	7.9±0.20	6.3±0.1	6.0–9.0
Coliforms (cfu/ mL)	1×10^{11}	1×10^{10}	1×10^8	≤1000
TP (mg/L)	52±3.0	29.1±0.20	1.4±0.24	0.5–1.5
SO_4^{2-} (mg/L)	1192±70.8	776.9±0.35	53.6±2.71	-
DO (% Saturation)	7±0.2	20.9±0.26	87.0±0.20	≥ 60
Temperature (°C)	22±1.5	37±0.5	37±0.5	< 35
COD (mg/L)	738±12.6	409.5±0.38	77.9±2.24	60–90

A_o: Acti-zyme-free MS effluent
A_1: MS effluent with Acti-zyme treatment
*Too many to count

Table 5.2: Overall summary on the effect of Acti-zyme on wastewater treatment

Parameter*	Type of wastewater			
	Wool scouring	**Dam water)**	**Piggery**	**MS**
	Cail et al. (1986)	**Tshuma (2010)**	**Dzvene (2013)**	**Current study**
BOD$_5$	62.4	96	58.1	92
COD	58.4	-	-	89.4
Total nitrogen	-	46	35.6	96
Total phosphates	-	67	-	97
Ammonia	-	53	48	-
Nitrates	-	80	-	-
E. coli	-	Positive	-	Positive
TSS	-	97.6	88.5	94
TDS	-	-	76.6	88.9
pH	-	6.8-7.4	7.82-7.84	6.8-7.4
Temperature	-	18–24°C	-	35°C
DO	-	100% increase	-	91% increase
Acti-zyme loading	1 % (v/v)	4 kg/day for 1 month in a 5 000m³ dam (0.024 kg/m³)	300m³ for 150 days (2 m³/day)	0.050 g/L
Retention time	207–211 days	60 days	150 days	40

Physico-chemical parameters measured in mg/L

Biogas generation

MS sludge, as with any other wastewater sludge, can serve as feedstock for biogas production using Acti-zyme as biocatalyst under anaerobic conditions (Duncan, 1970; Cail *et al.*, 1986). Although biogas has been produced from Acti-zyme catalysed wool scouring wastewater sludge by Cail *et al.* (1986) and from hog wastes by Duncan (1970), the effect of temperature on biogas production under mesophilic and thermophilic conditions still needs to be understood. Mesophilic conditions are typically employed for wet substrates with total suspended solids (TSS) of ≤15%, residence times of 60 to 95 days and complete mixing is required. Conversely, thermophilic conditions are required for wet substrates with TSS ≥20% at residence times of 9 to 45 days (Vindis *et al.*, 2009). Biogas and biosolid production from MS sludge, utilising Acti-zyme under anaerobic conditions can be represented by Reaction 1.

$$Sewage\ sludge \xrightarrow{Acti-zyme} CH_4 + CO_2 + H_2 + N_2 + H_2S + biosolids \tag{1}$$

Biogas production in a digester with substrate activated with Acti-zyme started immediately resulting in a low lag phase (Figure 5.2). The biogas obtained was characterised by CH_4 composition of 72% to 78%, CO_2 composition of 16% to 20% and trace gases composition of 8% to 12%. Biogas production increased concomitantly with Acti-zyme loading in the range 0–0.050 g/L for both the mesophilic (37°C) and thermophilic conditions (55°C) and all MS sludge loadings. Acti-zyme loadings of 0.050 g/L and MS sludge loading of 7.5 g/L.day were found to be optimal in terms of biodegradability of MS using Acti-zyme. MS sludge loading of 6.9–9.2 g/L. day has been recommended for optimal biogas production (Hesnawi & Mohamed, 2013). However, maximum biogas was achieved at mesophilic conditions with ~50% relative to the thermophilic conditions (Figure 5.2). This can be attributed to the Acti-zyme activity being optimal at temperatures ~37°C.

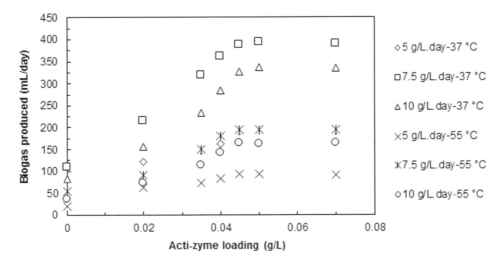

Figure 5.2: Effect of mesophilic and thermophilic temperature on biogas production at varying Acti-zyme and MSS loadings at a retention time of 40 days

Lower quantities of trace gases were noted in the digesters with Acti-zyme, due to the ability of Acti-zyme to inhibit the microbes responsible for production of traces like hydrogen sulphide (H_2S), hence improving the quality of the biogas (Table 5.3). The amount of biomethane generated in this study was also higher in comparison to previous studies (Table 5.4). A study by Duncan (1970) reported biogas characterised by 60% CH_4, from the digestion of hog waste mediated by 0.00625% of Acti-zyme for 50 days (35°C, hog waste loading = 0.75–0.99 kg/kg VSS). An investigation by Cail *et al.* (1986) reported biogas composition with 68% CH_4 from the digestion of wool scouring wastewater (35°C, wastewater loading= 0.5–1.5 L/day) using 1% (v/w) of Acti-zyme for 207 days (Table 5.4). These results suggest that biomethane production is also dependent on the nature of wastewater used. In comparison to the other reports, high biomethane quality in this study is attributed to higher Acti-zyme loading, which enhanced the MSS digestion (Table 5.3).

Table 5.3: Biogas composition from anaerobic digestion of MSS for systems with and without Acti-zyme

Gas	% (Acti-zyme)	% (without Acti-zyme)
CH_4	72–78	53–65
CO_2	16–20	22–27
Traces (H_2S, N_2, H_2)	5–9	8–12

CH_4 production was maximal at mesophilic conditions by more than 100% compared to thermophilic conditions. CH_4 yield was ~78% for the mesophilic conditions as compared to ~40% for thermophilic conditions at MS sludge loadings of 7.5 g/L.day. This indicated that mesophilic conditions are favourable for Acti-zyme catalysed biogas production from MS sludge as a value addition strategy.

Biosolids generation

Biosolids are generated as digestate during the Acti-zyme catalysed digestion of MS sludge (Figure 5.3). The biosolids generated were composed of 8.17±0.15%, 5.84±0.03% and 1.32±0.02% of nitrogen, phosphorous and potassium respectively. The NPK composition was higher, relative to the biosolids produced from MSS without aided digestion from Acti-zyme. The biosolids also contained copper (0.0073±0.0002%), iron (0.0087±0.0003%), calcium (0.0079±0.002%) and magnesium (0.016±0.0021%), which are micronutrients essential for plant growth. The biosolids obtained by this treatment method can be classified as high nitrogen content biosolids (Evanylo, 2009). The reduction of water content in the biosolids from 80% to 20% as well as the inhibitory effect of Acti-zyme for *E. coli* activity resulted in a significant decrease in *E. coli* content from 10^{12} to 10^6 cfu/L making the biosolids safe for application (Lang *et al.*, 2007). The amount of biosolids generated was lowest at 5–10 g/L.day at 37°C due to increased digestion associated with increased Acti-zyme activity which was absent at 55°C due to the high temperature (Figure 5.3). As the temperature increased from 37% to 55°C, the amount of biosolids produced showed an exponential decay trend for all the MSS loadings and Acti-zyme loadings.

Table 5.4: Summary of biogas generated in Acti-zyme catalysed systems and the process conditions involved

Type of wastewater	Acti-zyme loading	Retention time (days)	Organic loading rate	T (°C)	pH	Biogas production rate	Biomethane content	Reference
Wool scouring wastewater	1% (w/v)	207–211	0.75–0.99 kg/kg VSS	35°C	7.1–7.4	2.9–3.3 m^3/(m^3.day). 30% higher compared to an A_o system	68%	Cail et al. (1986)
Hog waste	0.00625% (w/v)	50	0.5–1.5 L/day	35°C	7.1–7.2		60% CH_4, 38% CO_2, 1% N_2, 1% water and H_2S traces	Duncan (1970)
MS sludge	0.050 g/L	40	7.5 g/L.day	37°C	7.3–7.5	400 mL/day	72–78% CH_4, 16–20% CO_2, and traces (H_2S, N_2, H_2)	Current study

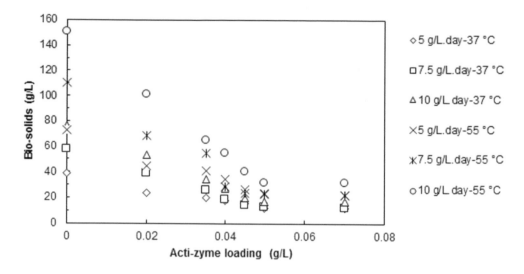

Figure 5.3: Effect of mesophilic and thermophilic conditions on biosolid production at varying Acti-zyme and MS sludge loadings and a retention time of 40 days

Techno-economic analysis

MS sludge, though a waste, if properly digested can help to meet the energy demands of a community utilising biocatalysts like Acti-zyme (Duncan, 1970; Cail *et al.*, 1986).

Techno-economic analysis for biogas and biosolid generation

Cost analysis of inputs for a conventional system

For a conventional MS treatment plant, aluminium sulphate is required as the coagulant for removal of the biocontaminants. The cost of treating the MS with the coagulant is pegged at $19 per ML of water (Dalton, 2008). For a plant capacity of 19 ML/day with an operating capacity of 60%, the total amount of aluminium sulphate required is $226 without value addition of sludge.

Cost analysis for an MS plant using Acti-zyme

The amount of Acti-zyme required at optimal process conditions of 0.050 g/L at a retention time of 40 days is calculated as follows:
Acti-zyme loading required 0.050 g/L at a retention time of 40 days for the 11,760, 000 L/day of MS is about 588 kg. This converts to $205 at Acti-zyme price of $0.35/kg (Powell and Lundy, 2007).

Amount of biogas generated

The biogas quantity generated was based on the optimum biomethane composition of 78%, which is the main component that is required in the biogas for electricity generation for an optimum MS loading of 7.5 g/L.day. Considering a 60% operating capacity and total solids content of 1143 mg/L, the amount of solids generated was 13,442 kg/day which converts to 12,770 kg/day of biogas was generated. If a cost price of $1.50/kg of the biogas is assumed, since this is the current selling price for gas in Zimbabwe based on the cost of liquid petroleum gas (http://www.lpgbusinessreview.com/2016/01/07/zimbabwe-africas-emerging-lpg-market/). This will result in income generated from the biogas per day to amount to $19,154.

Amount of biosolids generated

The biosolids produced are ~5% of the total amount of solids from this study. The amount of biosolids produced is 672 kg/day. If a selling price of $16/50 kg, which is currently the selling price for vermicompost in Zimbabwe (Zimbabwe Earthworm Farms, 2016), then $215 will be realised per day from the biofertilisers. The overall cost benefit analysis for using Acti-zyme in MST, co-digesting the MSS to biogas and biosolids is presented in Table 5.5.

Table 5.5: Cost benefit analysis for using Acti-zyme in MS treatment per day

	Product	Cost (US$)
Outputs	Biogas	19,154
	Biosolids	43
	Total	19,369
Less: Inputs	Acti-zyme	5
Net Benefit		**19,364**

If Acti-zyme is used in MS treatment harnessing biogas and biosolids, there is a net benefit of $19,364 per day, which translates to $5.6 million per annum and has the potential to make a significant input in the economy of Zimbabwe. The income generated from the biogas sales will contribute about 0.04% to the country's gross domestic product (GDP), which was US$13.7 billion according to Trading Economy (2014).

Financial benefit for electricity generation from biomethane

The potential for electricity generation from MS sludge biomethane is essential. Assumptions from Arthur and Brew-Hammond (2010) were adopted for determining the amount of electricity generated, that is, a biomethane heating value of 37.78 MJm^3, biogas engine efficiency of 29% and conversion factor 1kWh = 3.6 MJ (Arthur & Brew-Hammond, 2010).

The biomethane produced was calculated in accordance with the 78% composition in biogas and also by factoring in the density of the biogas, which is 1.15 kg/m^3 at standard conditions. The amount of biomethane generated was 456 m^3/day. Considering the energy efficiency of the biomethane and its conversion, the actual amount of bio-energy generated per day was 1 387 kWh. The financial benefit for the production of electricity from biomethane considered the investment required against the costs anticipated.

The amount of revenue that can therefore be realised from the electricity sales, given that the general price of electricity sales is $0.1/kWh in accordance with the Zimbabwe Electricity Authority tariffs, is therefore $13,679/ day. The investment that can be made for the MS plant is $5.5/kWh (Kottner, 2010), which translates to $7,602 per day and $2,219,950 per annum considering a plant operating capacity of 60% and 24 hours of operation. Several costs must be considered for biomethane electricity generation from MS sludge biomethane utilising Acti-zyme as digestion catalyst and these are indicated in Table 5.6. Maintenance cost is 2% of investment cost (Kottner, 2010), production cost is $0.035/kWh (Gebrezgabher *et al.*, 2010), insurance cost is 1% to 2% of investment cost (Kottner, 2010), labour cost is $22/hr (Kottner, 2010) and plant electricity usage is 7% of production cost (Kottner, 2010).

Table 5.6: Expected biogas plant operating costs per annum

Cost	Contribution	Value ($)
Maintenance	2% of investment	44,399
Production	$0.035/KWh	14,178
Insurance	1–2% of investment cost	44,399
Labour	$22/hr	154,176
Electricity	7% of production cost	992
Total Plant Costs		**258,145**

The final benefit for electricity generation from biomethane is indicated in Table 5.7. Investment has been taken to be for 10 years ($2,219,950×10 years = $22,199,504), while the other cash flows are up to 20 years.

Table 5.7: Cash flow for electricity generation from biomethane

Item	Cash flow ($)
Initial cash flow (Investment) (2 219 950.44*10)	22,199,501
Cash flow per year	
Electricity sold	3,994,286
Less: Total expected costs	258,145
Net Cash flow	**3,736,140**

Summary of economic factors

A summary of the economic factors is given in Table 5.8. The technology is economically feasible since a positive net present value (NPV) was derived. An internal rate of return (IRR) of 17.6% shows that the project adds value for stakeholders, since it is greater than the cost of capital, which was assumed to be 15%. The payback period of 5.9 years showed that it only takes approximately 5.9 years for the project to recoup its initial investment and the remaining 14 years are for profit generation. The breakeven point showed that for the project to start making a profit, only 183,059 kWh have to be produced with a sales value of $1.8 million. The MS treatment co-harnessing biogas is capable of producing 1387.33kWh/day indicating the potential high profitability of this approach.

Table 5.8: Summary of integrated MS management co-harnessing biogas and biosolids

Item	Value
Net present value	$1,186,239
Internal rate of return	17.6%
Payback period	5.9 years
Breakeven point in sales	183,059 KWh

CONCLUSION

Utilising Acti-zyme in integrated MS treatment co-harnessing biogas and biosolids is both environmentally and economically attractive. Acti-zyme effectively treated MSS, removing all the contaminants to meet the required guidelines for effluent disposal. MS treated with Acti-zyme showed >60% reduction in the MS contaminants at Acti-zyme loadings of 0.050 g/L and a retention time of 40 days. Anaerobic digestion of MS sludge utilising Acti-zyme at mesophilic conditions of 37°C promoted production of biomethane-rich biogas (78%), characterised by negligible hydrogen sulphate (H_2S) and nitrogen, and low CO_2 composition (16%). Additionally, biosolids rich in NPK nutrients are produced and can be utilised as biofertilisers.

Production of biogas from MS sludge using Acti-zyme as the digestion biocatalyst as a waste management technique is an economically viable process with a positive net present value (NPV), an internal rate of return (IRR) of 17.6% and a payback period of 5.9 years for an investment of $22 million over 20 years. The technology can be adopted as a value addition strategy for resource recovery from MS sludge for sustainable development in developing countries at the same time obtaining clean water for reuse.

REFERENCES

APHA. 2005. Standard Methods for the Examination of Water and Wastewater, 21st Edition, America Public Health Association, American Water Works, Association, Water Environment Federation, Washington, DC, USA.

Arthur, R and Brew-Hammond, A. 2010. Potential Biogas Production from MS Bio-solids: A Case Study of the MS Treatment Plant at Kwame Nkrumah University of Science and Technology, Ghana, *International Journal of Energy and Environment*, 1(6), 1009–1016.

Cail, RG, Barford, JP and Linchacz, R. 1986. Anaerobic Digestion of Wool Scouring Wastewater in Digester Operated Semi-Continuously for Biomass Retention, Agricultural Wastes, 18, 27–38.

Dalton, M. 2008. Potable Water Coagulant Trials Utilising Polyaluminium Chlorhydrate, 33rd Annual Old Water Industry Operations Workshop, Indoor Sports Centre, Carrara-Gold Coast, 3–5 June 2008.

Duncan, AC. 1970. Two Stage Anaerobic Digestion of Hog Wastes, Master of Science Thesis, University of British Columbia.

Dzvene, DK. 2013. An Investigation on the Effectiveness of Acti-zyme Bacteria to Treat Organic Waste at Triple C Pigs (Colcom), Bachelor of Science Honours Degree in Environmental Sciences and Technology, School of Agricultural Sciences and Technology, November 2013.

Evanylo, GK. 2009. Agricultural Land Application of Bio-solids in Virginia: Production and Characteristics of Bio-solids, Virginia Cooperative Extension, Publication, 425–301, 1–6.

Gebrezgabher, SA, Meuwissen, MPM, Prins, BAM and Oude Lausink, GJM. 2010. Economic Analysis of Anaerobic Digestion – A Case of Green Power Biogas Plant in the Netherlands, NJAS-Wageningen *Journal of Life Sciences*, 57(2), 109–115.

Hesnawi, RM and Mohamed, RA. 2013. Effect of Organic Waste Source on Methane Production during Thermophilic Digestion Process. *International Journal of Environmental Science and Development*, 4(4), 435–437.

Kottner, M. 2010. BioEnergy Farm, International Training Course, 2010.

Lai, TM, Shin, J and Hur, J. 2011. Estimating the Biodegradability of Treated MS Samples using Synchronous Fluorescence Spectra. *Sensors*, 11, 7382–7394.

Lang, NL, Bellet-Travers, MD and Smith, SR. 2007. Field Investigation on the Survival of Escherichia Coli and Presence of Other Enteric Micro-organisms in Bio-solids-Amended Agricultural Soil. *Journal of Applied Microbiology*, 103, 1868–1882.

LPG Business Review 2016. Available at: http://www.lpgbusinessreview.com/2016/01/07/zimbabwe-africas-emerging-lpg-market [Accessed 24 June 2016].

Manyuchi, MM, Ikhu-Omoregbe, DIO and Oyekola, OO. 2015. Acti-zyme Biochemical Properties: Potential for Use in Anaerobic MS Treatment Co-Generating Biogas. *Asian Journal of Science and Technology*, 6(3), 1152–1154.

Nhapi. I. 2009. The Water Situation in Harare, Zimbabwe: A Policy and Management Problem. *Water Policy*, 11, 221–235.

Nhapi, I and Gijzen, H. 2002. Wastewater Management in Zimbabwe, Sustainable Environmental Sanitation and Water Services, 28th Conference, Calcutta, India, p. 181–184.

Powell, B and Lundy, J. 2007. Acti-zyme Agricultural and Municipal Wastewater Treatment. Environment Depot Canada, p. 1–23.

Trading Economy. 2014. Zimbabwe GDP.

Tshuma, AC. 2010. Impact of Acti-zyme Compound on Water Quality along Mid-Mupfuure Catchment-Chegutu, BSc Honours Thesis, Geography and Environmental Studies.

Vindis, P, Marsec, B, Janzekovic, M and Cus, F. 2009. The Impact of Mesophilic and Thermophilic Anaerobic Digestion on Biogas Production. *Journal of Achievements in Materials and Manufacturing Engineering*, 36(2), 192–198.

Yang, K, Ji, B, Wang, H, Zhang, H and Zhang, Q. 2014. Bio-augmentation as a tool for improving the modified sequencing batch biofilm reactor. *Journal of Bioscience and Bioengineering*, 117(6), 763–768, 2014.

Zimbabwe Earthworm Farm. Available at: http://www.zimearthworm.com [Accessed 3 February 2016].

BIOGAS PRODUCTION FROM BLOOD AND RUMEN CONTENT OF SHEEP SLAUGHTERING WASTE UNDER AMBIENT CONDITIONS

R Niyobuhungiro and H von Blottnitz

Chemical Engineering Department, University of Cape Town,

Cape Town, 7700, South Africa

Corresponding author e-mail: harro.vonblottnitz@uct.ac.za

ABSTRACT

Slaughtering waste in poor peri-urban communities is often dumped at the roadside, creating health and safety risks. Such communities also generally lack clean and sustainable energy. Slaughtering waste can be treated to produce a clean-burning fuel, through biogas generation. While this technology is commonly used at commercial scale, it remains unknown whether it can be used in small anaerobic digesters at the side of the road, without features such as heating or stirring. Sheep slaughtering waste consisting mainly of rumen content and blood, collected from Nyanga, was digested in two 100-litre pre-fabricated fixed-dome type biogas digesters, operating at ambient temperatures (14–24°C), over a period of 90 days, fed daily at a rate of 80 g of solids. The biogas yield, pH, methane content and temperature were monitored throughout the period of the study. An average biogas production of 0.36 Lg^{-1} of volatile solids, equivalent to or 64 Lkg^{-1} of waste, was achieved. Ambient temperature affected the gas production to a limited extent. No short-term yield improvements resulted from doubling of the organic loading rate over 10-day periods but instead, the reactors became less stable. It is estimated that 100 kg of slaughtering waste dumped every day could generate enough biogas for seven vendors to be provided with thermal energy for their catering trades. Practical experiments are recommended to prove this technology in real settings.

Keywords: slaughtering waste, solid waste management, biogas, clean energy

INTRODUCTION

Energy is a key driving force for development. Its generation from fossil sources contributes to climate change by releasing greenhouse gas (GHG) into the atmosphere (Ahmad & Ansari, 2012). A global transition to renewable energy sources is underway.

Biochemical processes, like anaerobic digestion (AD), can produce clean energy in the form of biogas, which can be used for cooking and lighting (Ahmad & Ansari, 2012)

with significant potential also in South Africa (Stafford *et al.*, 2013). In addition, these AD processes could simultaneously resolve other ecological and agrochemical issues. The AD could also reduce the health and hygiene risks linked to slaughter waste by making the organics inaccessible to flies. For example, besides offering a waste disposal and treatment option, the AD of manure for biogas production does not reduce its value as a fertiliser, as available nitrogen and other substances remain in the treated sludge, which can be used in gardens (Alvarez & Lidén, 2008).

Due to the urbanisation in developing countries, many people from rural areas settle informally in the peripheries of the existing towns and cities. To cope with the urban life, they find themselves with no choice other than starting informal income-earning activities, which may often lead to the pollution and environmental compromise.

In informal settlements of Cape Town, one common activity is catering of cooked food and drinks. Cooked food includes meat, which comes from slaughtering activities. In one area close to the Nyanga transport interchange, approximately 40 animals (mainly sheep) are slaughtered daily, in addition to larger numbers of chicken. However, there are no slaughterhouses available in the areas. Caterers dump the slaughtering waste alongside the road. It is estimated that 2.5 kg of slaughtering waste (mainly rumen content) is thrown away for each slaughtered sheep, amounting to 100 kg of waste per day. The slaughtering waste creates significant public hygiene and health risks.

On the other hand, caterers and residents in these communities lack clean and sustainable energy. The use of waste and harvested wood in open fires in these communities is very inefficient. For instance, to pluck five chickens per day, a vendor was found to use 21.4 ± 2.6 kg of wood to produce the hot water needed, and consequently to work in a smoky environment with 3,500–6,500 μgm^{-3} of PM_{10} (particulate matter which are $\leq 10\mu$ in size) in the air (Niyobuhungiro, 2014). Before December 2014, the South African ambient standards for PM_{10} were 120 μgm^{-3} for 24 hours and 50 μgm^{-3} per year. From January 2015, the standards changed to 75 μgm^{-3} for 24 hours and 40 μgm^{-3} per year (DEA, 2014).

In the informal settlement of Nyanga, anaerobic digestion of slaughtering waste for biogas production might present one option for cleaner production. This would translate into a reduction in the dependence on wood and mitigate the atmospheric damages, including methane emissions from improper waste disposal (UNFCCC, 2010).

Even though the role of biogas production technology to utilise organic residues for energy production is known, the implementation of biogas technology is still a problem in many parts of the world (Tefera, 2009). However, biogas has been identified as one of the measures which has a large emission reduction potential for mitigating climate change and improving air quality (Shindell *et al.*, 2012).

Biogas technology might be one option for infrastructure provision to informal settlements in terms of solid waste management and renewable energy. In this regard, this chapter determines whether slaughterhouse waste can be used to produce biogas on a small scale; with 100-litre fixed dome digesters operating at ambient temperatures over 90 days. It presents a prediction of what could be achieved with scale-up from

100-litre digesters to 6 m³ units and briefly discusses the impact that this could have on the community.

MATERIALS AND METHODS

Sheep rumen content and a mixture of blood and rumen content were used as substrates. Slaughtering wastes were collected once in two weeks, packed in ziplock plastic bags, transported in cooler boxes and stored at -2°C in a freezer in the laboratory. The substrate was defrosted a day before feeding.

The characteristics of slaughtering waste from sheep and goats are reported in Table 6.1. The substrates were analysed for total solids (TS) and volatile solids (VS) using the drying oven for use at 103°C to 105°C (Laboratory & Scientific Equipment: Memmert, Lasec, South Africa) and a muffle furnace for use at 550°C (Carbolite [R], type: OAF10/2). The rumen content (grass matter) was determined to contain 19% total solids of which 92% VS. This makes this particular substrate fall in the category of medium solid AD system that contains 15 to 20% TS (Monnet, 2003).

Table 6.1: Substrate characteristics and feed rate

Substrate	Rumen content
% TS	19.1
% VS	92.1
Waste mass (gday⁻¹)	80
TS (gday⁻¹)	15.3
VS (gday⁻¹)	14.1
Total VS load (gL⁻¹ day)	0.23
Water added (mLday⁻¹)	200

The experiment was duplicated and conducted on two identical pre-fabricated fixed-dome type biogas digesters supplied by Agama Biogas: AD1 and AD2, nominally of 100-litre volume, of which 60 litres is active reaction volume and 40 litres is the expansion volume when gas is produced. Figures 6.1 (a) and (b) show the model and schematic configuration of the reactors used in the experiment. However, at the time when this experiment was carried out, the digesters had been in use for two years and they had not been emptied, which may have affected the normal active volume due to the solids accumulation. It is important to consider that different types of substrate had been used in these biodigesters before, depending on the purpose of the experiment. At least one of the prior experiments involved slaughtering wastes (Melamu *et al.*, 2012), so the culture had adapted to this kind of substrate.

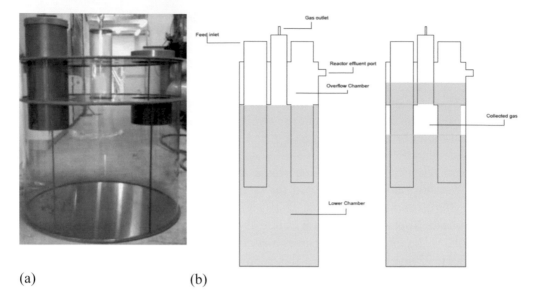

(a)   (b)

Figure 6.1: (a) Model, (b) Schematic configuration of the reactors used in the experiment

The biodigesters were generally subjected to the same conditions. In this experiment, the reactors were fed with 80 g of substrate every day except from days 58 to 67 and 81 to 90, where the feeding was doubled to see whether there is a difference in terms of gas production. The gas produced was measured and released every day when the reactors were fed. A sample of the sludge was taken for pH measurement.

The rumen content was the main feed, and it is noted to be rich in carbon. Once the pH inside the digesters dropped, the blood mixed with rumen content (richer in nitrogen) was added to raise the pH again. There was, however, no calculation on the quantity of the blood added as it was already mixed with the rumen content at collection.

The volume of biogas produced was measured using a water displacement method and the methane composition was checked using a gas analyser (Reiken-Kikki GX-2012A), which reads the CH_4 and O_2 in volume or molar percentages, and CO and H_2S in parts per million (ppm).

RESULTS AND DISCUSSION

Biogas and methane yields

This section will present and discuss the biogas production relative to active reactor volume, VS feed, and the specific methane production respectively. Figure 6.2 shows the total biogas production obtained in the experiment. Figure 6.3 expresses the biogas production in terms of volatile solids from the feed. Figure 6.4 shows the methane quantity in the biogas produced. This was calculated from the methane content of the biogas.

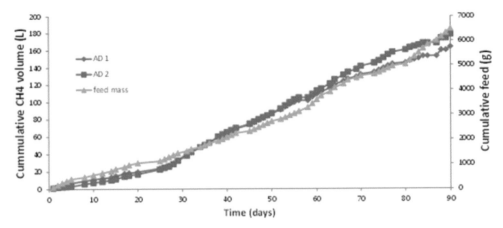

Figure 6.2: Cumulative biogas production relative to cumulative substrate addition

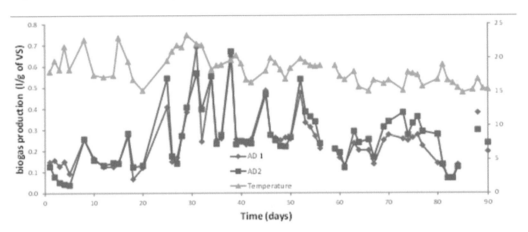

Figure 6.3: Daily biogas production and temperature trends

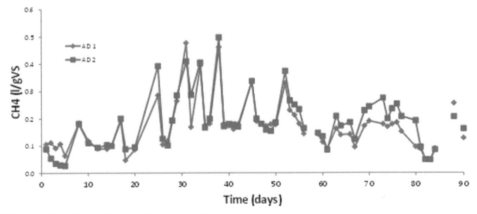

Figure 6.4: Specific methane yield

The average daily biogas volume harvested over the entire 90-day period was 3.95 and 4.12 L in reactors 1 and 2, respectively, at average methane concentrations of 69% and 71%. Biogas production rates were slower for the first 27 days. The average biogas production from days 28 to 90, normalised for the 60 litres of reactor volume was 0.06 $LL^{-1}day^{-1}$.

Considering the VS fed to the reactors every day, Figure 6.3 shows that before day 27, the biogas production was low (0.06 to 0.17 $Lg^{-1}VS$ for AD1, 0.04 to 0.17 $Lg^{-1}VS$ for AD2). This is due to the acclimation to the new substrate of the microorganisms. Before that day, the production was unstable and stabilised slightly afterwards. The biogas production achieved, reached 0.38 $Lg^{-1}VS$ fed for AD1 and 0.41$Lg^{-1}VS$ for AD2, both on day 29 and 0.40 $Lg^{-1}VS$ fed for AD1 and 0.48 $Lg^{-1}VS$ for AD2, both on day 64. However, due to the fluctuations of conditions, the average biogas production was observed to be 0.36 $Lg^{-1}VS$ or 4.5 L kg^{-1} of waste. The methane composition reached a maximum of 74% for AD2. The lowest percentage observed was 65% for AD1 and 68% for AD2.

The specific methane yield was seen to vary between 0.05 Lg^{-1} of VS to 0.34 $Lg^{1}VS$. In most cases, (from day 37 to day 75) the methane production was averaged to 0.23 $Lg^{-1}VS$ (Figure 6.4). Experimental methane yield from slaughterhouse waste was reported to be between 200 and 300 mLg^{-1} VS (Budiyono et al., 2011).

A decrease in the specific biogas production was observed for the days of doubled feed (58–67 and 81–90) essentially because the biogas production rate did not increase. This was done to see the effect of an increase in feed to the biogas production. It was seen that it somewhat destabilised the systems but that gas rates increased a few days afterwards. It is reported that the organic loading rate should be maintained the same to not disturb the micro-organism activity (Shayegan & Babaee, 2011).

Parameters that affected biogas production

In anaerobic digestion, biogas is produced from a complex process that involves a series of microbial metabolisms. The process can be divided into four main steps, hydrolysis, acidogenesis, acetogenesis, and methanogenesis (Fang, 2010).

There are many environmental and operational variables that determine the production of biogas in anaerobic digestion. The environmental requirement for the hydrolysis and acidogenesis phase is quite different from that of the acetogenesis and methanogenesis phase. Microorganisms in the methanogenesis phase have lower growth rates and are very sensitive to changes in environmental factors (Tefera, 2009).

Temperature and pH are two important process parameters that affect the biogas production. Their measured values in this experiment are discussed below.

Temperature

In our experiment, the temperature generally varied from 16°C to 19°C (ambient temperature), except for some particular days where it reached 23°C. This may explain the relative methane yield observed in our study. It was reported that a temperature fluctuation of even 1°F affects the biogas production (Tefera, 2009).

In this study, it is clear from Figure 6.3 that the temperature was unstable. From day 26 to 39, the temperature was between 18°C and 23°C. During this time the biogas production reached 41Lg^{-1} VS as the maximum temperature of 23°C for the whole experiment was reached. After that time, the temperature dropped, but this did not notably affect the gas production. Figure 6.5 shows the plot of the specific biogas production vs. temperature from day 27 onwards. The correlation coefficient shows that the correlation between the biogas production and the higher temperatures was weak under the conditions of this experiment.

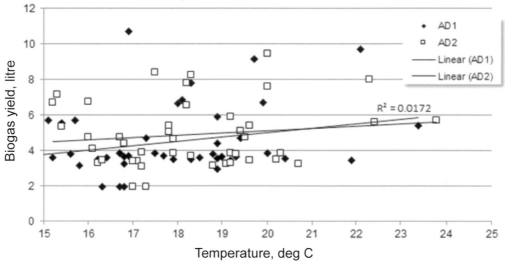

Figure 6.5: Daily biogas production vs. temperature

Tefera (2009) and Cuetos *et al.* (2009) confirmed that a rapid change in the temperature may result in a shutdown of the process or up to 30% loss in the biogas yield.

It is very important to acknowledge that normally, products from slaughterhouses, for example, rumen and intestinal content, have a high content of fat, lipids and protein that need longer hydraulic residence times (HRT) than biogas production from substrates like manure, silage and food waste. Even if the biomass type is easily degradable, the increase in temperature would reduce the time of the hydraulic retention and therefore increase the biogas yield (Rintala & Salminen, 2002; Cuetos *et al.*, 2009; Tefera, 2009).

Organic loading rate (OLR)

The optimum organic loading rate, which is a description of VS fed daily per unit of digester volume, depends on the temperature and the type and quantity of substrate (Fang, 2010). A feed rate exceeding that specific organic loading rate can result in incomplete degradation of the VS. This is an important parameter to be controlled and optimised for optimum gas production.

In the case of slaughtering waste used in this experiment, the organic loading rate was 0.23 gL^{-1}day^{-1} (Table 6.2). It should be increased to between 0.5 to 0.6 gL^{-1}day^{-1} for an optimum biogas production (CPCB, 2004; Budiyono *et al.*, 2012; Melamu *et al.*, 2011). The OLR should be maintained throughout the experiment. Changing the OLR in the middle of the experiment may destabilise the metabolism. In our experiment, when the OLR was doubled from day 59 to 67, it did not add anything in terms of biogas production. The second attempt, from days 81 to 90 resulted in destabilisation of metabolism.

pH

The pH is also an important parameter to get the stable operational conditions of the AD system because of the adverse effects that can occur when the pH fluctuates. Figure 6.6 shows that the pH varied between 7.8 and 8.4. Normally, in a properly operated anaerobic digestion, the pH should be between 6.8 and 7.2 (Monnet, 2003; Cioabla *et al.*, 2012; Melamu *et al.*, 2012). The formation of some by-products of the digestion affects the pH. If the substrate contains high amounts of proteins (e.g. slaughterhouse products), ammonia will be formed, which will raise the pH.

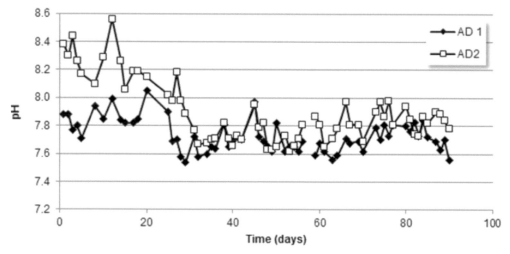

Figure 6.6: pH profile in the two digesters

In anaerobic digestion, microorganisms producing biogas are sensitive to the toxicity of ammonia, hydrogen sulphide and volatile fatty acids. Ammonia starts to be toxic to the microorganisms when the pH inside the biodigester is higher than 7 (Hashimoto, 1986; Yenigün & Demirel, 2013). The pH was well above 7 in this experiment, as shown in Figure 6.6. Ammonia levels were not measured.

Scale up estimates

One food-trading activity observed in Nyanga, is the slaughter, plucking and sale of chicken. To facilitate the plucking, the chicken is briefly dunked in boiling water, usually produced by inefficient fuel wood combustion, which is both costly and polluting. Table 6.2 presents an estimate of how much biogas would be needed to heat a 25-litre pot of water to boiling point for this activity. It further shows the quantity of rumen content that will suffice if the biogas yield 0.36 Lg⁻¹ VS could be achieved in an installation in the Nyanga setting.

Table 6.2 shows that 14.3 kg of waste should yield 0.91 m³ of biogas, sufficient for one vendor to accomplish the same activity of boiling enough water needed to pluck five chickens a day as observed. At an organic loading rate of 0.23 g L^{-1}day^{-1}, and a daily substrate feed of 2.5 kg VS, this would require a biodigester reactor volume of 11 m^3.

Table 6.2: Energy calculations for water heating

Quantity of water (kg)	25		
Heat used to boil 25.3 L of water (kj)	8705		
Heating value of biogas (kJm^{-3})	21300		
Theoretical biogas needed to generate the needed heat (m$_3$)	0.4	with the over design factor 45%	0.91
0.36 m^3 of biogas comes from 1 kg of VS; thus 0.91 m^3 would require	2.5	kg of VS	
TS needed for this (kg)	2.7		
Mass of substrate needed (kg)	14.3		

With 40 animals slaughtered daily and 2.5 kg of slaughtering waste (rumen content) from each animal thrown away, 100 kg/day of waste would be available. It means that enough biogas could be generated daily for seven chicken vendors, requiring a total reactor volume of 76 m^3. At the significantly higher OLR of 0.53 g L^{-1}day^{-1} and biomethane yield of 0.6 Lg^{-1}VS reported by Melamu *et al.* (2012) for co-digestion of blood with paper waste, a digester of half this size could produce twice as much biogas as estimated here. Fieldwork is recommended to test what performance can be achieved in real settings.

CONCLUSIONS

The objective of this experiment was to determine whether biogas could be produced from the anaerobic digestion of slaughtering waste (rumen content mixed with blood), in a non-heated, non-stirred anaerobic digester, and if so, how much. Over the 90 days of the experiment, biogas of good quality (70% methane) was robustly produced, averaging a yield of 0.36 Lg^{-1} VS fed once the digesters had stabilised after day 27. Higher ambient temperature was weakly correlated to higher gas yields.

At the performance criteria for digestion achieved in this experiment, the estimated 100 kg of slaughtering waste (dumped every day close to the Nyanga transport interchange in Cape Town), enough biogas could be generated daily for seven chicken vendors to be provided with energy for their activities.

Biogas installation at such sites of roadside slaughtering would provide the required solid waste management infrastructure and address air pollution from fuel wood usage. However, a scale-up to a demonstration facility is recommended to further evaluate feasibility.

ACKNOWLEDGEMENTS

We would like to express our gratitude to the Organisation for Women in Science for the Developing World for the fellowship for postgraduate studies. We acknowledge Nyanga community members who allowed the lead author to work with them as well as Miss Siphesile Tsotsa who helped with language translation. Special thanks go to Dr Thabi Melamu for assistance with the reactors and the original data analysis.

REFERENCES

Ahmad, J and Ansari, TA. 2012. Biogas from Slaughterhouse Waste: Towards an Energy Self-Sufficient Industry with Economical Analysis in India, J Microbial Biochem Technol , Harcourt Butler Technological Institute (HBTI), Kanpur, INDIA, http://dx.doi.org/10.4172/1948-5948.S12-001

Alvarez, R and Lidén, G. 2008. The effect of temperature variation on biomethanation at high altitude. *Bioresource Technology*, 99, 7278–7284.

Budiyono, B, Widiasa, IN, Johari, S and Sunarso, S. 2011. Study on Slaughterhouse Wastes Potency and Characteristic for Biogas Production. *International Journal of Waste Resources*, 1(2),4–7.

Central Pollution Control Board. 2004. Solid waste management in slaughter houses. Ministry of Environment and Forest. Government of India.

Cioabla, AE, Lonel, L, Dumitrel, G-A and Popesu, F. 2012. Comparative study on factors affecting anaerobic digestion of agricultural vegetal residues. *Biotechnology for Biofuels*, 5, 39.

Cuetos, MJ, Morán, A, Otero, M and Gómez, X. 2009. Anaerobic co-digestion of poultry blood with OFMSW: FTIR and TG-DTG study of process stabilization, *Environmental Technology*, 30, 6, 571–582, DOI: 10.1080/09593330902835730.

DEA (Department of Environmental Affairs). 2014. State of Air Report. Air Quality Standards and Objectives (Chapter 3).

Fang, C. 2010. Biogas production from food-processing industrial wastes by anaerobic digestion. DTU Environment, Department of Environmental Engineering, Technical University of Denmark, PhD thesis.

Hashimoto, AG. 1986. Ammonia inhibition of methanogenesis from cattle wastes. Agricultural Wastes, Elsevier.

Melamu, R, Bombile, J, Naik, L. and Von Blottnitz, H. 2012. Anaerobic co-digestion of paper sludge and blood waste: demonstrating a low-tech intervention for waste management in African cities.

Monnet, F. 2003. An introduction to anaerobic digestion of organic waste. Remade Scotland.

Niyobuhungiro, R. 2014. The applicability of a cleaner production approach in the case of Informal road side catering. Unpublished work.

Shayegan, J and Babaee, A. 2011. The Effect of Temperature on Methane Productivity of Municipal Solid Waste in Anaerobic Digester. *world congress on water, energy and climate.*

Stafford, W, Cohen, B, Pather-Elias, S, Von Blottnitz, H, Van Hille, R, Harrison, STL and Burton, SG. 2013. Technologies for recovery of energy from wastewaters: "Applicability and potential in South Africa". *Journal of Energy in Southern Africa*, 24(1), 15–26.

Rintala, JA and Salminen, EA. 2002. Semi-continuous anaerobic digestion of solid poultry slaughterhouse waste: effect of hydraulic retention time and loading, *Water Research*, 36, 3175–3182.

Shindell, D, Kuylenstierna, JCI, Vignati, E, Van Dingenen, R., Amann, M., Klimont, Z., Anenberg, SC., Muller, N., Janssens-Maenhout, G., Raes, F, Schwartz, J, Faluvegi, G, Pozzoli, L, Kupiainen, K, Höglund-Isaksson, L, Emberson, L, Streets, D, Ramanathan, V, Hicks, K, Oanh, NTK, Milly, G, Williams, M, Demkine, V and Fowler, D. 2012. Simultaneously mitigating near-term climate change and improving human health and food security. *Science*, 335, 183–189.

Tefera, TT. 2009. Potential for biogas production from slaughter houses residues in Bolivia (Systematic approach and solutions to problems related to biogas production at psychrophilic temperature). Stockholm, Sweden, master's dissertation.

UNFCCC (United Nations Framework Convention on Climate Change). 2010. Converting Waste from Slaughter House to Energy for productive use. Mitigation and Adaptation. Nairobi, Dagoretti, Kenya. Available at: http://unfccc.int/secretariat/momentum_for_change/items/7139.php [Accessed 29 May 2016]

Yenigun, O. and Demirel, B. 2013 Ammonia inhibition in anaerobic digestion: a review. *Process Biochemistry*, 48, 5, 901–911.

OPTIMISATION OF BIOGAS PRODUCTION BY CO-DIGESTION OF DOMESTIC ANIMAL WASTE

A Kazoka, JM Ndambuki and J Snyman

Tshwane University of Technology, Department of Civil Engineering, Private Bag X680,
Pretoria, 0001, South Africa
Corresponding author e-mail: SnymanJ@tut.ac.za

ABSTRACT

Energy plays an important role in the economic development of any country. The majority of the world's urban population depends on non-renewable energy sources generated from burning fossil fuels. This practice has a negative effect on the environment and should be avoided. The alternative would be the provision of renewable energy sources, which are adequate, affordable, efficient and reliable. This can be generated from various sources, such as solar, wind and organic waste. When organic waste undergoes a process called anaerobic digestion, it produces slurry and biogas which are both valuable products. Slurry is a good soil conditioner for crops and vegetables while biogas is used as a source of energy. Anaerobic processes take place in an airtight sealed container called a biogas digester. Research was conducted on the possibility of mixing three different livestock wastes in a single biogas digester (co-digestion) with a view to optimising biogas production. The livestock wastes used were cow dung, pig dung and chicken droppings collected from a rural village in Limpopo Province of South Africa. Five different mix ratios were determined based on daily available individual domestic animal waste from the study area and their carbon:nitrogen (C:N) ratio, which is one of the critical factors affecting biogas production from animal wastes. The study results show that the optimum biogas production was from a mix ratio of 1:1:1 (chicken droppings: pig dung: cow dung) digested at 32°C.

Keywords: biogas production, anaerobic digestion, waste type mix ratios, domestic animal wastes, waste optimisation

INTRODUCTION

Energy plays an important role in the economic development of every country. Therefore, provision of adequate, affordable, efficient and reliable energy services with minimum negative impact on the environment is crucial (Parawira, 2009). It is estimated that 3.2 billion people around the world lack access to modern methods of heating (Cheung, 2010). This is mainly the case in rural communities and is a significant barrier to economic development as reported by the United Nations Secretary-General's Advisory Group on Energy and Climate (United Nations, 2010). It is the aim of the Advisory Group that by 2030 everyone in developing countries should have access to modern energy services.

LITERATURE REVIEW

The world's largest energy supply is generated from non-renewable energy sources, which are fossil fuels (Venkateswara *et al*., 2010). However, this is increasingly becoming unsustainable (Venkateswara *et al.,* 2010) because fossil fuels, which are the main sources of national energy supply, cause ecological and environmental problems in addition to being non-renewable. Only 20% is generated from other sources such as nuclear, biomass, and renewable energy sources, such as solar, wind, hydro-electric and biogas.

The United Nations Environment Programme (UNEP, 2007) reported that most rural populations in developing countries prepare their meals on open fires fuelled by wood, coal, dry cow dung and charcoal. It is further reported by the UNEP (2007) that this practice adversely affects the health of rural communities that are exposed to indoor smoke generated during cooking. The cutting of trees for firewood and production of charcoal exhausts natural resources and promotes environmental degradation (Ukpabi, 2004). In order to provide rural communities with energy supply from the national grid, the available infrastructure requires expansion to meet the raising demand.

To mitigate this phenomenon, it is imperative that emerging affordable technologies that recognise the use of alternative renewable energy sources such as biogas be introduced (Maharjan, 2007), generated from organic waste (Amigun & Von Blottnitz, 2009). Biogas, which is a mixture of about 60% methane and 40% carbon dioxide, can be produced from organic wastes such as farm wastes under anaerobic digestion conditions through a single feedstock process or mixed feedstock process called co-digestion (Jingura & Matengaifa, 2009). Biogas ignites and burns with a blue flame and produces energy of calorific value in the range of 422 to 633 kilojoules (kJ) (Ahn *et al.,* 2009). This technology offers waste reduction while generating energy and can be used in rural areas for cooking (Austin & Blignaut, 2007; Amigun & Von Blottnitz, 2009).

Single feedstock anaerobic digestion involves only one type of organic waste being digested in an anaerobic digester while co-digestion is the simultaneous digestion of a homogenous mixture of at least two organic substrates in one digester (Jingura & Matengaifa, 2009). Co-digestion optimises biogas production and waste management. In this case, biogas production is optimised when substrates are correctly mixed (Mahnert & Linke, 2008). Research on co-digestion of farm wastes and municipal wastes to optimise biogas production has been conducted for over two decades. However, full implementation at household level has not been achieved due to the lack of technological dissemination programmes (Jingura & Matengaifa, 2009). Research on co-digestion indicates that the benefits in terms of waste management and biogas production are superior to those in the single digestion process (Tarek & Bashiti, 2010).

Principles of anaerobic digestion

The anaerobic digestion process occurs at three different conventional temperatures, by means of bacteria groupings known as psychrophilic, mesophilic and thermophilic (Adelekan & Bamgboye, 2009; Spuhler, 2010). The psychrophilic species functions at a temperature range from 4°C to 10°C with the optimal operating temperature ranging from 15°C to 18°C. These species specifically operate in cold temperatures. The mesophilic species can operate in temperatures as low as 10°C and their optimal operation temperature is between 30°C and 38°C. As for the thermophilic species, their optimal operation temperature is between 50°C and 52°C although they can operate in high temperatures up to 70°C (Adelekan & Bamgboye, 2009).

During anaerobic digestion, organic waste material is allowed to ferment in the absence of oxygen (Amigun & Von Blottnitz, 2009). The anaerobic digestion process works in two stages: the initial stage breaks down complex organics by acid-forming bacteria to simpler compounds; while the second stage converts volatile acids (such as acetic and propionic acid) by methane-producing bacteria to a mixture of methane and carbon dioxide. Anaerobic digestion is affected by factors such as temperature, the carbon:nitrogen (C:N) ratio, the pH of organic waste and the total solids (TS) of the waste sample.

Temperature

Temperature plays a critical role in biogas production, according to the Renewable Energy Institute (2005). The temperature within a biogas digester must be favourable for the specific type of anaerobic digestion bacteria to reach optimum biogas production (Adelekan & Bamgboye, 2009; Spuhler, 2010).

Carbon-nitrogen (C:N) ratio

The carbon-nitrogen (C:N) ratio of organic waste must fall between 20:1 and 30:1 for optimal biogas production according to the UNEP (2007). This is because if the C:N ratio is higher than 30:1, anaerobic bacteria will rapidly consume all the nitrogen to meet their protein requirements, thus starving some of the bacteria. Starved bacteria will in turn die, which consequently slows down the anaerobic digestion process. A lower C:N ratio than 20:1 will liberate some of the anaerobic bacteria, and accumulate in the form of ammonia which will increase the pH value of the mixture to more than 8.5 and slow down the biogas production (Adelekan & Bamgboye, 2009).

The ranges of the C:N ratios of chicken droppings, pig dung and cow dung are listed below (Zhu, 2007; Adelekan & Bamgboye, 2009; Belen *et al.,* 2011).

- Chicken droppings   : 15–24:1
- Pig dung   : 10–18:1
- Cow dung   : 10–15:1

pH and total solids (TS)

According to the UNEP (2007), anaerobic bacteria live optimally under conditions ranging from neutral to slightly alkaline. The pH value of organic waste, which falls in the range of 7 to 8.5, is ideal for optimal anaerobic digestion. If the pH of organic waste is acidic, it will have a toxic effect on the anaerobic bacteria. As for TS of organic waste, the range must be 7% to 10%. Therefore, the organic waste must be mixed with water, preferably grey water from the kitchen, to yield a TS percentage of the mixture within the range of 7% to 10% – preferably 8%.

The objective of this study was to determine a mix ratio of chicken droppings, pig dung and cow dung that optimises biogas production through anaerobic feedstock digestion under mesophilic conditions.

Types of biogas digesters

Different types of biogas digesters exist, which can be used on both large and small scale (Last, 2011). This study focused on small-scale digesters, which can be used at household level. Small-scale biogas digesters are airtight reactors filled with organic waste typically designed to produce biogas on a smaller scale, for example, at household or community level, especially in rural areas (Spuhler, 2010). Spuhler (2010) further points out that organic waste such as human excreta, crop waste, and kitchen waste can be directed to a single reactor in order to optimise biogas production by co-digestion. There are four types of small-scale biogas digesters according to German Technical Co-operation-Pure (2006) (Table 7.1). These are as follows:

1. Fixed-dome
2. Floating-drum
3. Bag-blanket, also known as the balloon digester
4. Earth-pit.

Each type of digester has its own advantages and disadvantages in terms of installation costs, maintenance costs, and functionality (Spuhler, 2010). Some of the advantages and disadvantages are as follows:

METHODOLOGY

The research was designed to collect both field and laboratory data. A series of 20 replicate laboratory experiments was conducted at different temperature ranges under mesophilic conditions. The aim of these experiments was to determine the optimal biogas production from individual and mixed domestic animal waste types. The experiments conducted in the laboratory were subjected to the critical factors which affect biogas production. The factors were specific to the study area and were as follows:

- Ambient temperature of the study area
- Available domestic animal waste
- Suitable type of biogas digester. This was established after an analysis of the economic standing of the community to determine which type of digester would be affordable in terms of capital and maintenance costs.

Table 7.1: Types of biogas digesters (German Technical Co-operation-Pure, 2006)

Type	Advantages	Disadvantages
1. Fixed-dome	• Low construction cost. • Long life cycle of up to 20 years because there are no moving parts or steel materials that could rust. • The digester can be constructed underground, which saves space and protects the digester from temperature fluctuations, especially at night and during winter periods. • Construction of a digester provides temporary employment for local people. • Construction materials are locally made except for cement, which can be sourced from the nearest town.	• The digester is vulnerable to gas leakages if not properly constructed. • Gas pressure fluctuates substantially depending on the volume of the gas stored.
2. Floating-drum	• A constant gas pressure is maintained due to vertical movements of the gas-holder. • The available gas volume can be determined by checking the level of the gas holder.	• High construction costs. • The gas holder is vulnerable to corrosion if not properly galvanised or if an un-galvanised drum is used, thus reducing the digester's life-span. • Requires regular maintenance, hence maintenance cost is high.
3. Bag-blanket	• The digester does not require highly skilled labour to construct. • The digester can be placed as close as possible to the kitchen; hence reducing the cost of gas piping required. • The digester can be successfully used in areas with a higher groundwater table where masonry or concrete walled digesters are unsuccessful.	• The digester is made of heavy-duty PVC plastic or rubber, which is not available in rural areas. • Repair work on torn plastic or rubber requires specialised plastic/rubber welding, which is not available in rural areas.
4. Earth-pit	• Low capital cost. • Temporary employment is created during the construction phase, and it can be built underground in low temperature areas to maintain the digester's operating temperature.	• Suitable for areas where the groundwater table is not high. • Requires special attention to the anchorage detail between the concrete ring and gas-holder to guarantee it to be leak free. • If steel sheeting is used as a gas holder, it corrodes in a short period; hence, reducing the life span of the digester.

Specific factors

Ambient temperature of the study area

Temperature plays an important role in biogas production according to the Renewable Energy Institute (2005). Therefore, it was important to determine the average ambient temperature of the study area. A maxima-minima thermometer was installed at the study area to record temperature readings daily for 26 consecutive days. From the maximum and minimum readings recorded, the average ambient temperature was calculated.

Available domestic animal waste

In order to determine how much animal waste was produced within the study area, the animals were gathered from their grazing areas every night and assembled in their respective kraals or sheds per household. All the waste produced overnight was collected and weighed. The total weight of the overnight waste production was divided by the total number of domestic animals that produced the waste, in order to determine production per animal. The total number of domestic animals (cow, pigs and chickens) per household was physically counted during field data collection.

The calculated waste production per animal was used to calculate the total expected waste production from the study area by multiplying waste production per animal by the total number of animals in the area. Thereafter, the calculated total waste production in the area was divided by the number of households to establish the daily available waste per household. This was feasible because the households were in close proximity to each other. This was done for 26 consecutive days, which is more than the seven-day minimum period recommended by the construction manual Modified GGC Model Biogas Plant for Pakistan 2009.

Twenty-six days were chosen in order to yield a wider range of results with which to compare and calculate an accurate average production per animal per day. This was done during the winter period, which is the worst period in terms of available grazing pasture for the animals. After collecting data for the available animal wastes, the mix ratios of the three domestic livestock wastes were determined based on the following fundamental factors:

1. Daily available fresh domestic livestock waste from the study area to effectively run a digester. This was determined by averaging the daily waste production of each domestic animal measured over a period of 26 days and multiplied by the number of domestic animals available per household.

2. Establishing the range of carbon:nitrogen ratio for each domestic animal waste. These were as follows (Zhu, 2007; Adelekan & Bamgboye, 2009; Belen *et al.*, 2011):

   - Chicken droppings   : 15–24:1

   - Pig dung   : 10–18:1

   - Cow dung   : 10–15:1

Taking the above two fundamental factors into account, different mix ratios were determined.

Suitable type of biogas digester

A biogas digester is an air-tight container that allows the decomposition of organic waste under anaerobic conditions to produce biogas (Renwick *et al.*, 2007). This process (Singh & Prerna, 2009) is a complex, natural, two-stage process of degradation of organic compounds through a variety of intermediates by the action of a consortium of micro-organisms (Amigun & Von Blottnitz, 2009).

Experimental materials

To measure the daily ambient temperature, a maxima-minima thermometer was used. For anaerobic digestion, fresh domestic animal wastes were used as substrates. The wastes were collected from the study area using 40-litre plastic containers. Three similar 40-litre containers were used as laboratory digesters and each one was correctly labelled. These plastic containers were put in a water bath, which was used to control digestion temperatures. The laboratory digesters were a simulation of a fixed-dome type digester installed underground. This type of biogas digester offers comparative advantages over others in relation to the study area as indicated, as it is cheaper to construct and maintain. The digesters were connected to 20-litre plastic containers used to collect biogas. The rate at which biogas was produced was measured in these plastic containers.

Experimental procedure

The laboratory experiments were conducted in two stages. The first stage involved biogas production for each single type of domestic animal waste. The second stage involved biogas production from mixed types of domestic animal wastes. The two stages of experiments were subjected to similar conditions and equipment. Thereafter, the results from the first stage of experiments were compared with the results of the second stage of experiments in order to determine the net increase in the optimum biogas production of the mixed type domestic animal wastes and hence, to determine the desired domestic animal waste mix ratio in this regard.

The average ambient temperature of the study area was determined by averaging the daily temperatures recorded over a period of 26 consecutive days. The moisture content of each fresh sample was measured in order to determine the additional quantity of mixing water required in each digester to achieve 8% total solids (TS) of the mixture. This was achieved by measuring the weight of each sample before and after drying in an oven for 24 hours at 110°C (Zhu, 2007). The difference between the mass of the fresh sample and dried sample was the quantity of moisture in the fresh sample. Thereafter, the laboratory digesters were assembled as shown in Figure 7.1.

Each fresh domestic animal waste was collected and weighed on a floor scale. Foreign objects, which were non-organic, such as stones and metals, detected in each sample were removed so that the sample only contained domestic animal waste. All three weighted samples were mixed in 40-litre plastic containers used as laboratory digesters. Each mixture of the samples was mixed with the correct amount of water to have 8% TS of the mixture. The containers shown in Figure 7.1 were connected to 20-litre gas collecting containers using 8 mm plastic tubes each fitted with two valves (flow control valve and sampling valve).

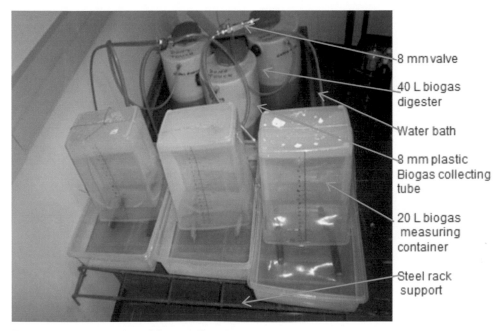

Figure 7.1: Laboratory biogas digesters

The digesters were put in a water bath fitted with a thermostat to ensure that the temperature was maintained according to the desired temperature level for that particular experiment. Measuring of biogas collected was stopped after 20 days because optimum biogas production was reached within the first 15 days from the start of the experiments. Replicate experiments for individual domestic animal waste types were conducted at four different temperature levels which were 28°C, 30°C, 32°C, and 34°C. This was done to determine the temperature level at which the optimum biogas production rate per individual domestic animal waste type was reached.

Thereafter, the second stage of replicate experiments was conducted for mixed domestic animal wastes at temperature levels 32°C and 34°C. The choice of 32°C and 34°C temperature levels for the mixed domestic animal wastes was based on the study area's average ambient temperature, results of individual domestic animal wastes and the optimum operating temperature range of mesophilic bacteria, which is 30°C to 38°C as established in previous research literature. The biogas produced was measured daily, and tested to check whether it would ignite (Figure 7.2). The results of optimum production from all mix ratios were compared with each other to establish the mix ratio with the highest production and to determine the increase in biogas production from the optimum production of an individual domestic animal waste.

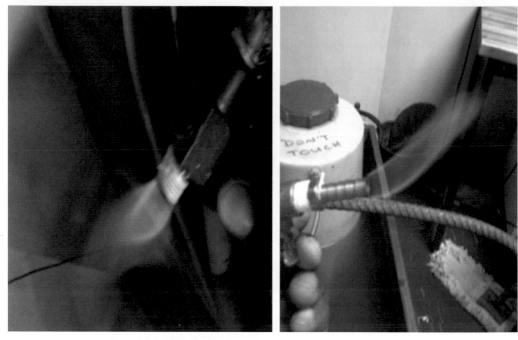

Figure 7.2: Igniting the biogas produced

RESULTS AND DISCUSSION

Average ambient temperature

The average minimum and maximum ambient temperatures of the study area were found to be 8°C and 23°C, respectively. The average of the two (minimum and maximum temperatures) was 16°C, which was 2°C lower than 18°C, which falls in the daily recommended temperature range for technically viable biogas production (German Technical Co-operation, 2006). To compensate for the 2°C difference, and to prevent temperature fluctuations in the biogas during the winter and at night, the digester should be constructed underground (Yetilmezsoy et al., 2007). A fixed-dome type biogas digester would be suitable.

The ambient temperature readings of the study area were taken in May and July, which are the coldest months of the year. Therefore, the calculated average ambient temperature (16°C) of the study area is considered the minimum average that the study area would experience in a year.

Determining the mix ratios

The mix ratios determined were based on the field investigation to establish the daily fresh available waste per household. It was established that 16.8 kg cow dung, 1.2 kg pig dung and 1.7 kg chicken droppings were produced daily. Chickens and pigs produced lower quantities of waste than cows daily. It was also established from the literature survey that chicken droppings have the highest C:N ratio ranges (15–24:1) compared to cow dung and pig dung. Therefore, the quantity of chicken droppings in all mix ratios was kept constant while pig dung and cow dung varied depending on the daily production. The mix ratios C : P : Cw (represented as C unit parts of chicken droppings : P unit parts of pig dung : Cw unit parts of cow dung) were as follows: 1:6:21; 1:6:10; 1:3:21; 1:1:2; and 1:1:1.

To make it easier when measuring quantities of waste to be mixed, all of the mix ratios figures except for 1:1:2 and 1:1:1 were directly used as mass in kg of the fresh samples. The total mass of mix ratios 1:1:2 and 1:1:1 used were 20 kg and 15 kg respectively. This was because to have 8% TS, the total combined mass with mixing water would occupy less than 50% of the 40-litre laboratory plastic container used as digester if a direct conversion of mix ratios to mass in kg as for the other mix ratios was done. Thus, it would not be effectively digested to produce the desired results. For anaerobic digestion to effectively take place, the organic waste and mixing water in the digester must occupy more than 50% of the total digester volume according to the construction manual, Modified GGC Model Biogas Plant for Pakistan 2009. The same quantities of samples and their respective total mass that were used to mix with water for each mix ratio as indicated in Table 7.2 were used at each different temperature level.

Table 7.2: Different mix ratios of domestic animal waste samples

Mix ratio number	Mix ratio parts per unit measure	Mass of sample as per mix ratio	Total mass of sample used per digester
C : P : Cw	1 : 6 : 21	1 kg . 6 kg : 21 kg	28 kg
C : P : Cw	1 : 6 : 10	1 kg : 6 kg : 10 kg	17 kg
C : P : Cw	1 : 3 : 21	1 kg : 3 kg : 21 kg	25 kg
C : P : Cw	1 : 1 : 2	5 kg : 5 kg : 10 kg	20 kg
C : P : Cw	1 : 1 : 1	5 kg : 5 kg : 5 kg	15 kg

The domestic wastes in Table 7.2 were freshly collected from the study area. However, they did not have enough moisture for effective anaerobic digestion. Therefore, they had to be mixed with water and the mixture in each digester was more than 50% minimum required for the digester to be effective.

Determining the moisture contents of each sample

The results of the moisture content of each sample collected for experiments were converted to percentage mass of the sample itself, so that the additional mixing water required for each sample would be calculated based on the percentage of available moisture in each waste sample. Table 7.3 shows the moisture content of each waste type and the calculated percentage of water required for mixing to have a homogeneous mix with 8% TS.

Table 7.3: Moisture content and mixing water per waste sample

Type of domestic animal waste sample	Moisture content	Percentage of water required for mixing/unit sample
Cow dung	47%	45%
Pig dung	47%	45%
Chicken droppings	53%	39%

The quantities of mixing water were calculated in kg using the percentages of water required per sample in Table 7.3. The mass of each sample used was 20 kg and experiments were done at 28°C, 30°C, 32°C and 34°C temperature levels for each waste type. Results are shown in Figures 7.3, 7.4, 7.5 and 7.6.

Biogas production analysis

Individual type of domestic animal wastes

The biogas production results for experiments done for each domestic animal waste type are presented as follows:

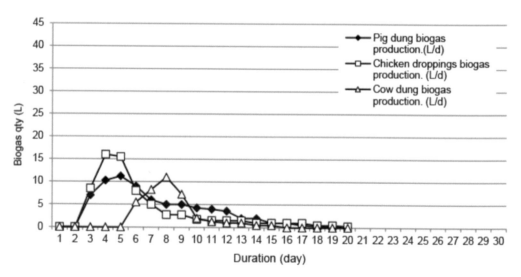

Figure 7.3: Daily biogas production at 28°C per type of animal waste

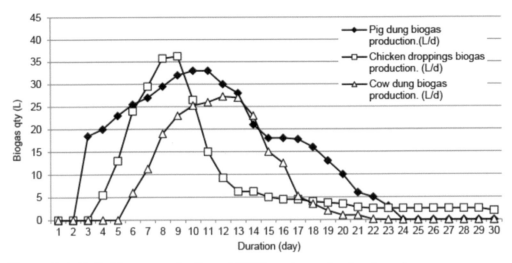

Figure 7.4: Daily biogas production at 30°C per type of animal waste

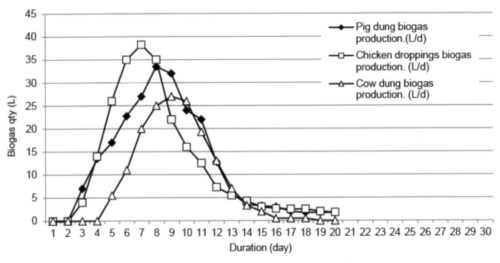

Figure 7.5: Daily biogas production at 32°C per type of animal waste

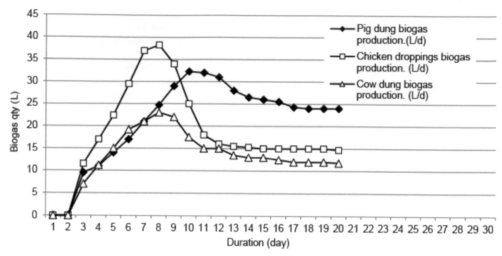

Figure 7.6: Daily biogas production at 34°C per type of animal waste

The results of optimum biogas production rate per domestic animal waste at each temperature level were analysed and compared to each other. The optimum production rate per unit (1 kg) was calculated and results are shown in Table 7.4.

Table 7.4: Optimum biogas production per 1 kg of animal waste

Type of domestic animal waste	Optimum production [L/d]			
	at 28°C	at 30°C	at 32°C	at 34°C
Chicken droppings	1.10	2.42	2.55	2.55
Pig dung	0.75	1.97	2.23	2.15
Cow dung	0.73	1.82	1.80	1.53

It is clear from the results shown in Figures 7.3, 7.4, 7.5 and 7.6 and Table 7.4 that chicken droppings produced the highest biogas production rate. Furthermore, the highest production rates were reached at temperature levels 32°C and 34°C. Therefore, these results were taken into consideration when establishing mix ratios and the temperature levels at which experiments for mix ratios were conducted.

Mix type of domestic animal wastes

The biogas production results for experiments with different mix ratios conducted at 32°C and 34°C were analysed and compered to the results of biogas production of individual type of domestic animal wastes.

Biogas production at 32°C

There was no biogas produced on day one and day two as shown in Figure 7.7. The first biogas production recorded from all mix ratios was on day three. The optimum production rates from mix ratio 1:1:1 and 1:1:2 occurred on the 11th and 10th day, respectively. The other mix ratios, which are 1:6:21, 1:3:21 and 1:6:10 reached optimum production rate on the 14th day. The decline in biogas production from all mix ratios showed a similar steep trend until production stabilised.

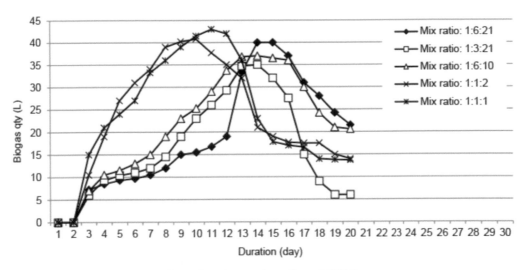

Figure 7.7: Biogas production of various mix ratios at 32°C

The highest cumulative biogas produced over a 20-day period as illustrated Table 7.5 was derived from the 1:1:1 mix ratio followed by the 1:1:2 mix ratio. However, these production rates are based on different sample weights.

Table 7.5: Biogas production from all mixed samples at 32°C

Mix ratio	Weight of mixed sample [kg]	Optimum biogas production [l/d]	Total cumulative biogas production [l]
1:6:21	28	40.0	378.55
1:3:21	25	35.0	325.65
1:6:10	17	37.0	429.10
1:1:2	20	40.8	467.95
1:1:1	15	43.0	473.65

To calculate the optimum biogas production rate per day, and cumulative biogas produced over a 20-day period per sample of 1 kg of mixed domestic animal waste, both the optimum and cumulative productions in Table 7.5 were divided by the respective total mass of the mixed sample. The results are presented in Table 7.6.

Table 7.6: Biogas production per kg of mixed sample at 32°C

Mix ratio	Weight of mixed sample [kg]	Optimum biogas production [l/d]	cumulative biogas production [l]
1:6:21	1	1.43	13.52
1:3:21	1	1.40	13.01
1:6:10	1	2.18	25.24
1:1:2	1	2.04	23.4
1:1:1	1	2.87	31.6

Biogas production at 34°C

Experiments for biogas production at 34°C were conducted only for the first three mix ratios with highest optimum production rate per 1 kg sample at 32°C. These were the 1:1:1, 1:1:2 and 1:6:10 mix ratios. This was because the aim of the study was to investigate one mix ratio that would optimise biogas production rate. Therefore, it was only necessary to check the productuion rates of the three mix ratios at 34°C temperature level. The results are shown in Figure 7.8.

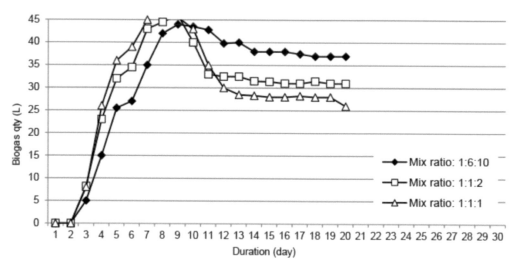

Figure 7.8: Biogas production of various mix ratios at 34°C

Biogas production started on the third day. The mix ratios 1:1:1 and 1:1:2 produced more biogas on the first day than the mix ratio 1:6:10. The highest optimum biogas production rate was derived from 1:1:1 and 1:1:2 mix ratios as demonstrated in Figure 7.8 and Table 7.7, which was reached on the 8th and 9th days, respectively. The second highest optimum production rate was derived from the 1:6:10 mix ratio, which occurred on the 9th day.

Table 7.7: Biogas production from mixed samples at 34°C

Mix ratio	Weight of mixed sample [kg]	Optimum biogas production [l/d]	Total cumulative biogas production [l]
1 : 1 : 1	15	46.0	576.55
1 : 1 : 2	20	46.0	587.60
1 : 6 : 10	17	44.0	622.10

The mix ratio waste samples had different weights. Therefore, to calculate optimum production rates per kg of each mixed sample, the total weight of the mixed sample was divided into the sample's optimum production rate. The results are shown in Table 7.8. These results show that mix ratio 1:1:1 still had the highest production rate per kg while mix ratio 1:6:10 was the second highest.

Table 7.8: Biogas production from 1 kg mixed samples at 34°C

Mix ratio	Weight of mixed sample [kg]	Optimum biogas production [l/d]	Total cumulative biogas production [l]
1 : 1 : 1	1	3.1	38.4
1 : 1 : 2	1	2.3	29.38
1 : 6 : 10	1	2.6	36.6

The two sets of results in Table 7.6 and Table 7.8 show that mix ratio 1:1:1 had the highest optimum production rate at both temperature levels with the highest occurring at 32°C temperature level.

Comparison of biogas production of individual and mixed type animal wastes

Individual domestic animal wastes

The highest optimum biogas production rate results obtained from individual types of domestic animal wastes (especially chicken droppings) were compared with mixed domestic animal wastes to establish whether mixed animal wastes optimised biogas production rates and by how much. The individual optimum biogas production rates were first compared by calculating the percentage increases in biogas production rates from a lower temperature range to a higher temperature range. Equation (1) and data in Table 7.4 were used for these calculations.

$$\left(\frac{(a-b)}{c}\right) \text{ x } 100\% = x\% \tag{1}$$

Where:

a = Optimum biogas production per kg of waste at a higher temperature range.
b = Optimum biogas production per kg of the same waste at a lower temperature range.
c = Optimum biogas production per kg of waste at a lower temperature range.
x = Percentage increase in biogas production.

The results obtained are as follows:

- **Chicken droppings**
  - i. 28°C to 30°C: $\left(\frac{(2.42\,L-1.1\,L)}{1.1\,L}\right)$ x 100% = 120%
  - ii. 30°C to 32°C:L$\left(\frac{(2.55\,l-2.42\,L)}{2.42\,L}\right)$ x 100% = 5.4%
  - iii. 32°C to 34°C: $\left(\frac{(2.55\,L-2.55\,L)}{2.55\,L}\right)$ x 100% = 0%
- **Pig dung**
  - i. 28°C to 30°C: $\left(\frac{(1.97\,L-0.75\,L)}{0.75\,L}\right)$ x 100% = 163%
  - ii. 30°C to 32°C: $\left(\frac{(2.23\,L-1.97\,L)}{1.97\,L}\right)$ x 100% = 13.2%
  - iii. 32°C to 34°C: $\left(\frac{(2.15\,L-2.23\,L)}{2.23\,L}\right)$ x 100% = -3.6%
- **Cow dung**
  - i. 28°C to 30°C: $\left(\frac{(1.82\,L-0.73\,L)}{0.73\,L}\right)$ x 100% = 149%
  - ii. 30°C to 32°C: $\left(\frac{(1.8\,L-1.82\,L)}{1.82\,L}\right)$ x 100% = -1%
  - iii. 32°C to 34°C: $\left(\frac{(1.53\,L-1.8\,L)}{1.8\,L}\right)$ x 100% = -15%

The results indicate an increase in the optimum production rate of biogas for the experiments that were conducted at 28°C and 30°C of more than 100% for all the types of domestic animal waste. The increase in production for the experiments conducted at 30°C and 32°C for the chicken droppings was 5.4% and 13.2% for the pig dung. On the other hand, the optimum biogas production rate derived from the cow dung dropped slightly by 1%. The results from experiments conducted at 34°C using the chicken droppings, revealed no increase in the optimum biogas production rate compared with those conducted at 32°C. However, the results from the pig dung and the cow dung dropped by 3.6% and 15%, respectively.

The cumulative biogas production from all the waste samples was higher for the experiments conducted at 32°C and 34°C than for those experiments carried out at 28°C and 30°C. In addition to the higher cumulative biogas production at 34°C, after reaching optimum levels, the biogas production gradually dropped for four days and subsequently, maintained production between 12 L to 24 L per day per sample which was higher than those experiments conducted at 28°C and 30°C. This proved that mesophilic bacteria lived longer at temperatures of 32°C and 34°C than at 28°C and 30°C.

Mixed domestic animal waste at 32°C

The highest optimum biogas production was derived from a mix ratio 1:1:1 followed by 1:6:10. The difference in the optimum production rate between the two mix ratios calculated from data in Table 7.6 was 31.7%. The percentage difference in optimum production rate between the two mix ratios at the same temperature level was calculated using Equation (1) and data in Table 7.6. Note that 'b' and 'c' values are replaced with optimum biogas production value of a mix ratio 1:6:10. The results are as follows:

- Optimum production rate from mix ratio 1:1:1 = 2.87 L/kg
- Optimum production rate from mix ratio 1:6:10 = 2.18 L/kg
- 1:6:10 to 1:1:1: $\left(\frac{(2.87\ L - 2.18\ L)}{2.18\ L}\right)$ x 100% = 31.7% (increase in production rate)

This is an indication that having more chicken droppings and pig dung in a mixture increases the production rate of biogas than having more cow dung.

Mixed domestic animal waste at 34°C

The results of the mixed domestic animal waste experiments conducted at 34°C indicate that optimum biogas production was derived from the mix ratio 1:1:1. This further proves that mix ratio 1:1:1 has highest biogas production rate even at different temperatures. The optimum results from a mix ratio 1:1:1 demonstrates an increase of 8% from the results of the same mix ratio at 32°C. The data used to calculate for percentage increase is from Tables 7.6 and 7.7 using Equation (1). Below are the results:

- Optimum production at 32°C temperature   = 2.87 l/kg
- Optimum production at 34°C temperature   = 3.1 l/kg
- 32°C to 34°C: $\left(\frac{(3.1\ L - 2.87\ L)}{2.87\ L}\right)$ x 100% = 8% (increase in production rate)

Comparison of individual and mixed domestic animal wastes

The highest optimum biogas production rate results derived from the individual types of domestic animal waste were compared with those obtained from mixed domestic animal waste. The percentage increases were calculated as follows:

- Optimum production at 32°C temperature from chicken droppings = 2.55 L/kg/d.
- Optimum production at 32°C temperature from 1:1:1 mix ratio = 2.87 L/kg/d.
- Percentage increase in production rate: $\left(\frac{(2.87\ L - 2.55\ L)}{2.55\ L}\right)$ x 100% = 12.5%

- Optimum production at 34°C temperature from chicken droppings = 2.55 L/kg/d.
- Optimum production at 34°C temperature from 1:1:1 mix ratio = 3.1 L/kg/d.
- Percentage increase in production rate: $\left(\frac{(3.1\ L - 2.55\ L)}{2.55\ L}\right)$ x 100% = 21.6%

The results indicate that the production rate from mixed wastes at 32°C increased by 12.5% compared with that of the highest recorded individual types of waste, that is, from chicken droppings. The increase in the optimum production rate of biogas from the mixed domestic animal waste at 34°C was 21.6% higher than that of the highest optimum production from the individual types of domestic animal waste at the same temperature level.

CONCLUSIONS

In this study, the results of biogas production from the three mixed domestic livestock waste, namely cow dung, pig dung and chicken droppings, have been presented. The results show that mixing the three domestic livestock wastes in equal quantities (1:1:1) can optimise the biogas production rate. Therefore, to improve livestock waste management and prevent the use of firewood, which encourages environmental degradation, it is recommended that the communities adopt biogas technology as their source of energy. The livestock waste produced daily from the study area was adequate to run a 4 m^3 biogas digester effectively and produce 52.1 litres of biogas per household per day, which is adequate to completely replace firewood as a source of energy for cooking. The households are in close proximity and those households with less livestock waste produced daily, can access livestock waste from their neighbours with excess livestock waste. Therefore, each household can have a 4 m^3 biogas digester in which the daily livestock waste produced can be mixed in equal quantities to produce biogas, which can be used for cooking. This will not only improve waste management in the area, but also provide affordable and reliable clean energy.

REFERENCES

Adelekan, BA and Bamgboye, AI. 2009. Effect of mixing ratio of slurry on biogas productivity of major farm animal waste types. *Journal of Applied Biosciences,* 22, 1333 –1343.

Ahn, HK, Smith, MC, Kondrad, SL and White, JW. 2009. Evaluation of biogas potential by dry anaerobic digestion of Switchgrass-Animal manual mixtures. *Journal of applied biochem biotechnol,* 160 (2010), 965–975.

Amigun, B and Von Blottnitz, H. 2009. Capital cost prediction for biogas installation in Africa: Lang factor approach. Department of Chemical Engineering, Environmental & Process Systems Engineering Research Group, University of Cape Town, South Africa. Available at: http://www.interscience.wiley. com). DOI 10.1002/ep.10341. [Accessed 5 June 2011].

Austin, G and Blignaut, J. 2007. *South Africa national rural domestic biogas feasibility assessment.* Available at: http://www.agam.co.za [Accessed 10 June 2011].

Belen, P, Ponsa, S, Gea, T and Sanchez, A. 2011. Determining C:N ratios for typical organic wastes using biodegradable fractions. *National Center for Biotechnology Information Journal,* 85, 653–659. Available at: http://www.elsevier.com/locate/chemosphere [Accessed 2 July 2016].

Cheung, PK. 2010. *Biogas plants improve life for rural families in Nepal. Biogas is a healthier and greener option than cooking over an open fire.* Available at: http://www.dw-world.de/dw/article/0,6143866,00. html [Accessed 27 October 2010].

German Technical Co-operation-Pure 2006. *Feasibility study report on biogas from poultry droppings. Sanitation, 28 December 2005. Promotion of Biogas Production.* Available at: http://www.lged-0rg/ archive-file/FILD_VISIT_REPORT_III.pdf [Accessed 2 January 2011].

Jingura, RM and Matengaifa, R. 2009. Renewable and sustainable energy reviews. Optimisation of biogas production by anaerobic digestion for sustainable energy development in Zimbabwe. *Renewable and Sustainable Energy Reviews,* 13, 1116–1120.

Last, S. 2011. *Anaerobic digestion.* Biogas digester: Heart of anaerobic digestion process. Available at: http://www.anaerobic-digestion.com [Accessed 16 February 2011].

Mahajan, N. 2007. *Future generation energy: move over wind and solar energy, cow dung is here to stay.* Available at: http://www.ecofriend.com/mooove-over-wind-and-solar-energy-cows-poop-is-here-to-stay.html [Accessed 28 October 2010].

Mahnert, P and Linke, B. 2008. *Kinetic study of biogas production from energy crops and animal waste slurry: effect of organic loading rate and reactor size.* Available at: http://www.informationworld.com [Accessed 8 June 2011].

Modified GGC Model Biogas Plant for Pakistan. 2009. *Construction manual*. Available at: http://www. snvworld.org/../ construction_manual_modified_GGC_model_biogas_plant_for_pakistan_2009.pdf [Accessed 10 July 2011].

Parawira, W. 2009. Biogas technology in Sub-Saharan Africa: status, prospects and constraints. *Reviews Environmental Science and Biotechnology*, 8, 187–200.

Renewable Energy Institute. 2005. *Anaerobic digester: how anaerobic digesters work, and how anaerobic digesters produce bio-methane. The green of all biofuels*. Available at: http://www.anaerobicDigester. com [Accessed 4 February 2011]

Renwick, M, Subedi, PS and Hutton, G. 2007. Biogas for better life: An African initiative. *Draft final report on cost benefit analysis of national and regional integrated biogas and sanitation programmes in sub-Sahara Africa*. Dutch Ministry of Foreign Affairs *to* Winrock International. Available at: http:// www.winrock.org/clean_energy/files/ biogas_for_better_life_an_african_initiative.pdf [Accessed 4 February 2011].

Singh, SP and Prerna, P. 2009. Renewable and sustainable energy reviews. Reviews of recent advances in anaerobic packed-bed biogas reactors. *Renewable and Sustainable Energy Reviews,* 13, 1569–1575.

Spuhler, D. 2010. *Sustainable sanitation and water management. Anaerobic digestion (small scale)*. Available at: http://www.sswm.info/category/implementation-tools/wastewater-treatment/hardwear [Accessed 14 June 2011].

Tarek, A and Bashiti, EI. 2010. *Biogas production by co-digestion of animal manure and olive oil wastes. Journal of Al-Azhar University to Gaza*, released on the occasion of the international conference on basic and applied science (ICBAS2010), 12, 27–30.

Ukpabi, C. 2004. *Biogas for better life an African Initiative*. Available at: http://www.biogasafrica.org [Accessed 16 February 2011].

UN (United Nations). 2010. *Report by United Nations Secretary General Advisory Group on Energy and Climate Change. By 2030 All People to have access to modern energy*. Available at http://www.vdc-tonckehocevar.com/energy/united-nations-2030-all-people-have-access.com [Accessed 25 January 2011].

UNEP (United Nations Environment Programme). 2007. Case study summary on 2007 Ashden Awards; SKG Sangha India*: The Ashden Awards for sustainable energy.* Available at http://www.skgsangha. org [Accessed 25 January 2011].

Venkateswara Rao, P, Saroj, SB, Ranjan, D and Srikanth, M. 2010. Biogas generation potential by anaerobic digestion for sustainable energy development in India. *Renewable and Sustainable Energy Reviews*, 14, 2086–2094

Yetilmezsoy, K, Kocak-Enturk E and Ozturk, M. 2007. A small-scale biogas digester model for hen manure treatment. Evaluation and suggestions. *Fresenius environmental bulletin by PSP*. Available at: http:// www.yildiz.edu.tr/~yetilmez/1st%20page%20Jour.%20Article6.pdf [Accessed 29 May 2016]

Zhu, N. 2007. Effect of low initial C:N ratio on aerobic composting of swine manure with rice straw. *Journal of Bioresource Technology,* 98, 9–13.

INTEGRATED BIOREMEDIATION AND BENEFICIATION OF BIOBASED WASTE STREAMS

SH Rose,[1] **L Warburg,**[1] **M Le Roes-Hill,**[2] **N Khan,**[2] **B Pletschke,**[3] **and WH van Zyl**[1]

[1] Department of Microbiology, Stellenbosch University, Private Bag X1, Matieland, 7602, South Africa

[2] Biocatalysis and Technical Biology Research Group, Institute of Biomedical and Microbial Biotechnology, Cape Peninsula University of Technology, PO Box 1907, Bellville, 7535, South Africa

[3] Department of Biochemistry and Microbiology, Rhodes University, PO Box 94, Grahamstown, 6140, South Africa

Corresponding author e-mail: WHVZ@sun.ac.za

ABSTRACT

Potato processing results in the generation of different types of waste. In this study, novel applications were investigated for the possible use of potato waste streams. The wild type *Trametes pubescens* strain was able to produce laccase (2 U mL^{-1}) when cultivated on potato waste. The activity increased to 4.4 U mL^{-1}, when red grape waste was added to the cultivation medium. Recombinant *Aspergillus niger* D15[pGTP], D15[eg2], D15[man1] and D15[xyn2] strains were cultivated on potato waste and potato wastewater, and evaluated for the production of industrially relevant enzymes. Endoglucanase, endomannanase and endoxylanase activity levels of 38, 20 and 114 U mL^{-1}, respectively, were obtained after three days of cultivation on a potato waste (mixed solids). The use of wastewater did not have a negative impact on the levels of enzyme activity. In addition, an autoselective amylolytic *Saccharomyces cerevisiae* Y294[AG] strain was constructed and evaluated for the production of bioethanol from potato waste. The addition of a commercial enzyme cocktail resulted in an increase in ethanol production from 5.2 to 6.2, and 3.7 to 4.3 g L^{-1} on heated and raw starch, respectively. This study demonstrated that a starch-rich waste can be utilised for the production of bioproducts such as enzymes and bioethanol, and highlights the potential role that alternative biotechnological processes can add to the bioremediation and beneficiation of organic-rich waste streams.

Keywords: potato waste biomass, microbial conversion, heterologous enzymes, genetic engineering, consolidated bioprocessing, sustainable fuels and chemicals

INTRODUCTION

Lignocellulose, sugars and phenolic compounds are commonly found in the solid waste and wastewater fractions of vegetable and fruit processing industries. In South Africa, potatoes are cultivated on approximately 50,000 hectares and yield more than 2 million tons per annum (Potatoes South Africa, 2015). The gross value of potato production accounts for about 43% of major vegetables, 15% of horticultural products and 4% of total agricultural production. On average, domestic potato farmers harvest about R1.6 billion worth of potatoes a year (DAFF, 2013).

Potatoes are a versatile vegetable with more than 50% of the potato crop being processed into food products, animal feed, starch and seed tubers for the next season's crop (Potato processing and uses, 2013). The dry matter and starch content of South African potato varieties range from 12 to 24% and 8 to 16%, respectively (Leighton *et al.*, 2009). Starch is composed of two related polymers in different proportions according to its source, amylose (16–30%) and amylopectin (65–85%). Amylose is a linear polymer consisting of α-1,4 linked glucose units. Amylopectin is a large highly branched polymer of glucose with α-1,6 bonds at the branch points (Hashem & Darwish, 2010). Potato pulp is reported to contain 55% galactose, 9% arabinose, 17% galacturonic acid and 1.4% rhamnose – no fucose, xylose or mannose (Thomassen *et al.*, 2011; Van Dyk *et al.*, 2013).

Globally, the consumption of potato as food is shifting from fresh potatoes to processed potato products (including frozen potatoes). Potato processing produces solid wastes consisting of peels and whole or cut potatoes that are discarded for reasons such as size or blemishes. The waste and discarded potatoes can be collected at different stages during the process (Figure 8.1A), constituting different waste streams. Losses can constitute 15% to 40% depending on the peeling processes used (Schieber *et al.*, 2001). The solid waste is used for animal feed, whereas the liquid waste is disposed of through the general sewerage system (Figure 8.1B). Catarino *et al.* (2007) described the general flow process for a crispy chips industry, which included the following: potato discharge, washing, peeling, slicing, blanching, frying, quality control and packaging – and reported the successful recovery of by-products.

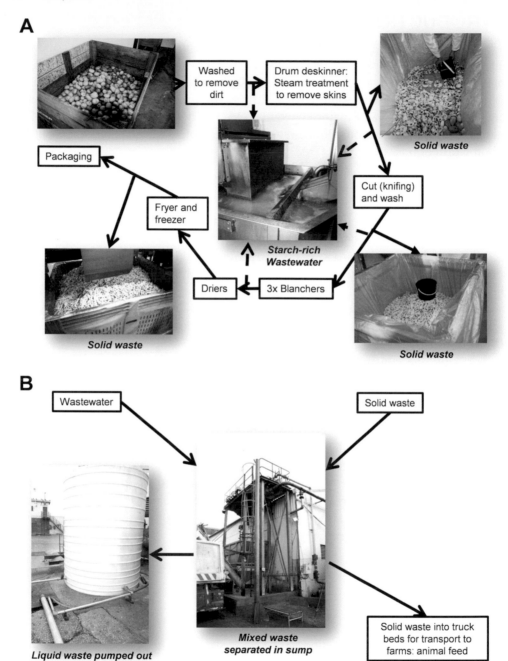

Figure 8.1: (A).The different types of solid and liquid waste generated during potato processing. The solid wastes used in this study are indicated. (B) Wastewater and solid waste are separated in the sump; liquid waste is pumped out (e.g. general sewerage), and the solid waste is used for animal feed (San Martin *et al.,* 2016)

Food processing industries also generate large volumes of wastewater due to the washing of produce and equipment. All wastewater must meet certain legal requirements before being discharged into rivers or used for irrigation. In South Africa, water with a chemical oxygen demand (COD) of less than 400 mg L^{-1} can be used for irrigation at volumes of up to 500 m³ (Khan *et al.*, 2015). Irrigation volumes may not exceed 50 m³ on any given day when the COD is between 400 mg L^{-1} and 5000 mg L^{-1}. The average COD of wastewater in the juicing and canning industries is often as high as 10,000 mg L^{-1}, therefore extensive treatment is required before the water can be discharged into rivers (Van Dyk *et al.*, 2013). Management of both the solid waste and the wastewater fractions is a priority for the food processing industry due to health concerns. Wastewater treatment is costly; therefore, value-added products need to be generated from the waste to offset the cost of the treatment.

Aspergillus niger and *Saccharomyces cerevisiae* are food-grade eukaryotic microbes that have a long history with the food and beverage industries. Fungal species such as *Phanerochaete chrysosporium*, *Aspergillus* spp. and *Trametes* spp. can degrade and utilise a wide range of phenolic compounds (García-García *et al.*, 1997) making them ideal for the removal of complex organic matter in waste streams. The filamentous fungus, *A. niger*, is the organism of choice for industrial enzyme production due to the copious amounts (grams per litre) of native enzymes secreted for the hydrolysis of starch, pectin and cellulose (De Vries & Visser, 2001; Jin *et al.*, 1998). Various *A. niger* strains have been constructed that produce cellulases, xylanases, mannanases (Rose & Van Zyl, 2002; Rose & Van Zyl, 2008; Van Zyl *et al.*, 2009), and laccases (Bohlin *et al.*, 2006) with cultivation on glucose. The *A. niger* strains also demonstrated heterologous enzyme production when cultivated on lignocellulosic waste streams after the hexose sugars (in sugar cane bagasse and northern spruce hydrolysates) were fermented to ethanol with industrial yeast strains (Alriksson *et al.*, 2009; Cavka *et al.*, 2011; Cavka *et al.*, 2014).

A key objective of this study was to find novel processes for the integrated bioremediation and beneficiation of potato wastes. Proposed processes include the production of enzymes of commercial value using a wild type *Trametes pubescens* strain and recombinant *A. niger* strains cultivated on biobased waste streams (potato waste), as well as production of bioethanol from potato waste using an amylolytic *S. cerevisiae* strain.

MATERIALS AND METHODS

Biomass waste preparation (carbohydrate source)

A potato processing facility in the Western Cape supplied five different potato wastes (kouebelt detector [KD], langebelt detector [LD], slivers, skins and starch). The KD waste consisted of chips with defects collected after the steam peeler and oil fryer treatment. LD waste consisted of chips with defects collected after the steam peeler treatment, while the sliver waste consisted of small potato off-cuts. Slivers and skins were collected after the steam peeler treatment, and the potato mixed solids (PMS) were a mixture of the different types of potato wastes obtained from the removal trucks. The potato waste was homogenised and stored at -20°C until use. The red grape waste (RGW) used in this study was obtained from a cellar in the Stellenbosch area (Western Cape, South Africa) and stored at -20°C without further processing.

Fungal strains and media composition

The relevant fungal genotypes of the strains used in this study are listed in Table 8.1. All media components were of analytical grade and sourced from Merck unless stated otherwise. The *T. pubescens* culture was maintained on Trametes Defined Media (TDM) agar plates (10 g L^1 glucose, 5.25 g L^{-1} peptone, 2 g L^{-1} KH$_2$PO$_4$, 0.5 g L^{-1} MgSO$_4$·7H$_2$0, 0.1 g L^{-1} CaCl$_2$, 0.3 g L^{-1} NaCl, 12 g L^{-1} agar and 10 mL trace element solution consisting of 0.6 g L^{-1} FeSO$_4$, 0.03 g L^{-1} CuSO$_4$, 0.07 g L^{-1} ZnCl$_2$, 0.34 g L^{-1} MnSO$_4$, 0.19 g L^{-1} CoCl$_2$, 0.002 g L^{-1} NiCl$_2$ and 0.62 g L^{-1} (NH$_4$)$_2$MoO$_4$. Cultures were incubated at 28°C for five to seven days and plates were stored at 4°C. One agar plate (covered in *T. pubescens*) was homogenised in 100 mL sterile dH$_2$O using a Waring blender. One mL homogenate was used to inoculate 20 mL culture medium [20% wet weight (KD or PMS) v^{-1} with 1% (w v^{-1}) RGW]. Cultivation took place at 28°C at an agitation speed of 200 rpm. Supernatant samples were obtained after 3, 5, 7 and 10 days of cultivation.

The *A. niger* D15[eg2] and D15[man1] strains are similar to the strains referred to in Rose and Van Zyl (2008), but contain a functional *pyrG* gene. The *A. niger* strains were maintained on spore plates [2 g L^{-1} neopeptone (Difco), 1 g L^{-1} yeast extract, 0.4 g L^{-1} MgSO$_4$·7H$_2$0, 10 g L^{-1} glucose, 2 g L^{-1} casamino acids, 20 mL AspA (300 g L^{-1} NaNO$_3$, 26 g L^{-1} KCl, 76 g L^{-1} KH$_2$PO$_4$, pH 6) and 1 mL L^{-1} 1000 × trace elements]. Trace elements were prepared according to Punt and Van den Hondel (1992).

Table 8.1: Fungal strains used in the study

Fungal strains	Relevant genotype	Enzyme[#]	Reference
T. pubescens	wild type	none	CBS 696.94
A. niger			
D15[pGTP]	*PyrG gpd$_P$-glaA$_T$*	none	Chimphango *et al.* (2012)
D15[eg2]PyrG*	*PyrG gpd$_P$-eg2-glaA$_T$*	Eg2	This study
D15[man1]PyrG*	*PyrG gpd$_P$-man1-glaA$_T$*	Man1	This study
D15[xyn2]PyrG*	*PyrG gpd$_P$-xyn2-glaA$_T$*	Xyn2	Rose and Van Zyl (2008)
S. cerevisiae			
Y294[BBH1]	*URA3 ENO1$_P$-ENO1$_T$*	none	Viktor *et al.* (2013)
Y294[AmyA-GlaA]	*URA3 ENO1$_P$-amyA-ENO1$_T$*	AmyA	Viktor *et al.* (2013)
	URA3 ENO1$_P$-glaA-ENO1$_T$	GlaA	
Y294[AG]	*URA3 ENO1$_P$-amyA-ENO1$_T$*	AmyA	This study
	ENO1$_P$-glaA-ENO1$_T$ FUR1::LEU2	GlaA	

Heterologous enzymes expressed
* *Strains are referred to as D15[eg2], D15 [man1] and D15[xyn2], respectively, in the text.*

The *A. niger* strains were cultivated on potato waste medium (PWM), which was prepared by autoclaving the traditional medium [10 g L^{-1} yeast extract, 0.8 g L^{-1} MgSO$_4$·7 H$_2$O and 4 g L^{-1} casamino acids] containing the potato waste (10 and 20% wet weight), followed by the addition of 40 mL AspA and 2 mL trace elements. The final volume in the 125 mL Erlenmeyer flasks was approximately 21 mL. Strains were inoculated to a spore count of 5×10^5 spores per mL. Cultivation took place at 30°C at an agitation speed of 200 rpm.

The *S. cerevisiae* Y294[AmyA-GlaA] strain had previously been constructed by Viktor *et al.* (2013), expressing the *amyA* and *glaA* genes originating from *Aspergillus tubingensis*. The native *FUR1* gene was disrupted using the *LEU2* marker gene to generate an autoselective strain, *S. cerevisiae* Y294[AG], in accordance to La Grange *et al.* (2001). Strains were maintained on YPD agar plates (10 g L^{-1} yeast extract, 20 g L^{-1} peptone, 20 g L^{-1} glucose and 15 g L^{-1} agar). Pre-cultures of the reference strain, *S. cerevisiae* Y294[BBH1], and the *S. cerevisiae* Y294[AG] strain were prepared in YPD medium and cultivated to a cell density of 3×10^8 cells per mL. A 10% (v v^{-1}) inoculum was used for the fermentation studies using 100 mL yeast potato waste medium [YPWM: 10 g L^{-1} yeast extract, 20 g L^{-1} peptone, 200 g L^{1} (wet weight) PMS]. The YPWM was heated in a microwave for two minutes (to 60°C) and cooled to room temperature prior to the addition of 2 mg L^{-1} lactoside 247 (Lallemand, USA) and 20 µL Stargen 002 (Genencor, The Netherlands). Agitation (200 rpm) and incubation were performed on a magnetic multi-stirrer at 37°C, with regular sampling through a syringe needle pierced through the rubber stopper.

Enzyme assays and analytical tests

Ostazin brilliant red hydroxyethyl cellulose (OBR-HEC), OBR locust bean gum (OBR-LBG) and Remazol brilliant blue birchwood xylan (RBB-xylan) were prepared according to Biely *et al.* (1985). Synthetic complete (SC) agar plates [6.7 g L^{-1} yeast nitrogen base without amino acids (Difco Laboratories, USA), 100 g L^{-1} glucose], containing 0.1% (w v^{-1}) OBR-HEC, OBR-LBG or RBB-xylan, respectively, were prepared.

The extracellular endoglucanase, endomannanase and endoxylanase activity was determined using the reducing sugar method described by Bailey *et al.* (1992) with 0.1% (w v^{-1}) lichenan (Sigma-Aldrich, Germany), 0.25% (w v^{-1}) locust bean gum (Sigma-Aldrich, Germany), and 1% (w v^{-1}) beechwood xylan (Roth, Germany) as substrates. Glucose, mannose and xylose were used as standards to determine the endoglucanase, endomannanase and endoxylanase activities respectively. All assays were performed at 50°C and pH5 (0.5 M citrate phosphate buffer). The supernatant was appropriately diluted prior to the five-minute incubation with the substrates. The reactions were terminated by the addition of the 3,5–dinitro-salicylic acid (DNS, Sigma-Aldrich, Germany) solution and boiled for 15 minutes at 100°C to assist in colour development. The colorimetric changes were measured spectrophotometrically at 540 nm with an X-MARK™ microtitre plate reader (Biorad, Hercules, CA, USA). The activities were expressed in U mL^{-1}, where one unit of enzyme activity was defined as the amount of enzyme required to release one µmol of glucose, mannose or xylose per minute.

Laccase activity was measured using 0.5 mM ABTS [(2,2'-Azino-bis(3-ethyl benzothiazoline-6-sulfonic acid) diammonium salt, Sigma Aldrich, Germany] dissolved in 100 mM sodium acetate buffer (pH 5), as substrate. Fifty μL of culture supernatant was incubated with 150 μL of ABTS solution for three minutes at 22°C. The colorimetric changes were measured spectrophotometrically at 420 nm ($\varepsilon = 36\ 000\ M\ cm^{-1}$) in the microtitre plate reader. One unit of enzyme activity was defined as the amount of enzyme required to oxidise one μmol of substrate at ambient temperature ($22 \pm 2°C$). *Trametes versicolor* laccase (Sigma-Aldrich, Germany) was used as a positive control for all laccase assays.

Quantification of metabolites

The glucose, ethanol, glycerol and acetic acid concentrations were determined using high performance liquid chromatography (HPLC), with a Waters 717 injector (Milford, MA, USA) and Agilent 1100 pump (Palo Alto, CA, USA). The compounds were separated on an Aminex HPX-87H column (Bio-Rad, Richmond, CA), at a column temperature of 60°C with 5 mM H_2SO_4 as mobile phase at a flow rate of 0.6 mL min^{-1} and subsequently detected with a Waters 410 refractive index detector. Concentration ranges used for standard curves varied from 0.5 g L^{-1} to 50 g L^{-1}.

RESULTS AND DISCUSSION

Starch-rich waste and waste streams from potato processing facilities are potentially excellent carbohydrate resources that can be utilised for the cultivation of various microorganisms. In this study, this biobased waste was used for the cultivation of a wild type fungal strain with the ability to produce the industrially relevant laccase enzyme.

The *T. pubescens* strain was cultured in a medium containing potato waste (KD) with and without the addition of RGW. The extracellular laccase activity was monitored over time (Figure 8.2). No laccase activity was detected with cultivation on the RGW without KD in seven days (data not shown), probably due to the insufficient carbohydrate source hampering the growth of the fungus. Similar levels of laccase activity were observed with cultivation on KD and PMS. The addition of 1% (w v^{-1}) RGW initially contributed to an increase in laccase activity, which is not surprising as the RGW contains high levels of lignin due to the presence of stalks and pips. However, the laccase activity in general remained low with less than 5 U mL^{-1} produced during the time of cultivation. An increase in RGW did not result in a significant change in the levels of activity obtained (data not shown). While phenolic compounds are known to induce laccase production in *T. pubescens*, higher concentrations of RGW appeared to delay the onset of laccase production. The results do not correspond with previous studies that indicated an increased laccase production by *T. pubescens* in winery wastewater with high phenolic concentrations (Strong & Burgess, 2008; Strong, 2011). The use of different strains and potato waste as a carbon source could account for the discrepancy observed, which may be related to the induction of the native laccase encoding genes.

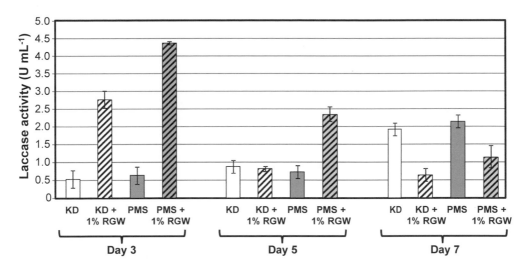

Figure 8.2: Laccase production by *T. pubescens* over a period of seven days on 20% w v^{-1} KD, KD supplemented with 1% (w v^{-1}) RGW, PMS and PMS supplemented with 1% (w v^{-1}) RGW. Error bars represent standard deviations from triplicate experiments

The low levels of enzyme production by the *T. pubescens* wild type strain were most probably related to a lack of induction of the native laccase genes. Therefore, the majority of this study focused on recombinant *A. niger* strains (Table 8.1) as a source of industrially relevant enzymes. The *A. niger* strains used in this study were based on previously constructed strains (Rose & Van Zyl 2002; 2008) and are known to produce high levels of extracellular enzyme activity. The use of the glyceraldehyde 3-phosphate gene (*GPD*) promoter results in the constitutive expression of the foreign genes (Table 8.1), circumventing the problems associated with induction of gene expression. Heterologous endoglucanase, endomannanase and endoxylanase enzyme activities were confirmed on OBR-HEC, OBRLBG and RBBxylan plates, respectively (Figure 8.3A, B and C). Clearing zones around the colonies are an indication of the extracellular enzyme activity. Although for the reference strain, *A. niger* D15[pGTP], also produces endoglucanase, endomannanase and endoxylanase activities, no activity was detected on the plates due to the presence of the 10% glucose resulting in catabolite repression of the native genes.

The KD, LD, skins, slivers and starch potato wastes were used to prepare PWM containing 10% wet weight biomass. The extracellular enzyme activities of the *A. niger* D15[pGTP], [eg2], [man1] and [xynB] strains were determined on day three and day six in PWM10% (Figure 8.3D). The endoglucanase, endomannanase and endoxylanase activity produced by the reference strain, *A. niger* D15[pGTP], never exceeded 3.5 U mL^{-1} and was therefore omitted from the graphs. In most cases, an increase in activity was observed on day six (data not shown), but the increase in activity did not justify an additional three days of cultivation.

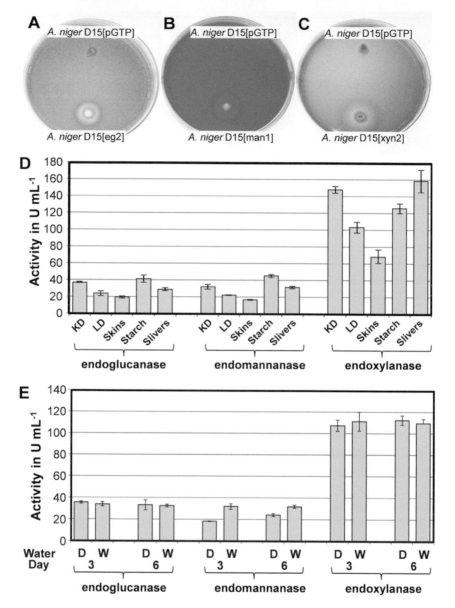

Figure 8.3: The recombinant *A. niger* D15[pGTP], D15[eg2], D15[man1] and D15[xyn2]
strains were transferred to SC plates containing (A) OBR-HEC for
endoglucanase activities, (B) OBR-LBG for endomannanase activities, and
(C) RBB-Birchwood xylan for endoxylanase activities. The clearing zones
around the colonies are representative of the different activities. The *A. niger*
D15[GTP], D15[eg2], D15[man1] and D15[xyn2] strains were cultivated at
30°C in PWM–10% using 10% (w v^{-1}) (D) separated and (E) mixed potato
waste prepared in [D] distilled and [W] waste water. The extracellular
endoglucanase, endomannanase and endoxylanase enzyme activities
were determined after three days and after six days of the cultivation. Error
bars represent standard deviations from triplicate experiments

An increase in potato waste content resulted in a slight increase in activity. For all three strains, the activity observed using skins as biomass source, was much lower than with the other carbohydrate sources. The skins are removed from the potatoes after a heat treatment, resulting in little starch material remaining attached to the skins. The skins contain primarily cellulose, which is more difficult for the strains to utilise as a carbohydrate source. The recombinant *A. niger* strains were evaluated on 10% (w v⁻¹) PMS waste prepared in distilled water and wastewater to simulate an industrial process. The water had little effect on the fungal growth or the levels of activity (Figure 8.3E) produced by the strains indicating that the wastewater from the potato industry can be used for the cultivation of *A. niger* for enzyme production. Similar results were obtained with the PMS waste compared to that obtained with cultivation on the KD, starch and slivers (Figure 8.3D) suggesting that the potato waste can be used without separation of the individual components. The concentration of the potato waste was limited to 10%, as higher levels of viscosity hampered mixing and aeration.

The recombinant *S. cerevisiae* Y294[AmyA-GlaA] strain had previously been constructed with the ability to hydrolyse raw starch through the heterologous expression of the *A. tubingensis* α-amylase and glucoamylase encoding genes (Viktor *et al.*, 2013). The use of this amylolytic strain would require less of the commercial amylase cocktail (such as Stargen 002) for optimal starch hydrolysis. However, the *S. cerevisiae* Y294[AmyA-GlaA] strain cannot be used in an industrial process due to the instability of the episomal vector in non-selective industrial media. The strain was therefore converted to a semi-industrial strain through genetic engineering involving the disruption of the *FUR1* gene (Kern *et al.*, 1990) resulting in the autoselective *S. cerevisiae* Y294[AG] strain (Table 8.1).

The *S. cerevisiae* Y294[AG] and *S. cerevisiae* Y294[BBH1] strains were cultivated anaerobically in YPWM, containing 200 g L⁻¹ PMS as sole carbohydrate source. Samples were taken daily and the extracellular metabolite concentrations determined (Figure 8.4). The fermentations were completed after four days of cultivation. Ethanol production was accompanied by the simultaneous production of acetic acid and glycerol at lower concentrations (Figure 8.4A to F). The initial ethanol and glycerol concentrations observed at T_0 are either due to carry over from the 10% inoculum or due to the conversion of the potato waste by natural organisms (no sterilisation of the medium prior to inoculation). The increase in acetic acid (Figure 8.4D) concentration over time is indicative of the unsterile cultivation conditions combined with the inability of the *S. cerevisiae* Y294[BBH1] strain to produce sufficient ethanol (in the absence of Stargen 002) to control the bacterial growth.

A significant difference in ethanol yield could be observed between the *S. cerevisiae* Y294[AG] and *S. cerevisiae* Y294[BBH1] strains with cultivation on the heated PMS but not the raw PMS. Heating of the PMS presumably made the starch more accessible for the Stargen 002 enzymes and the recombinant enzymes produced by the *S. cerevisiae* Y294[AG] strain. The addition of Stargen 002 improved the starch conversion on heated PMS for both the *S. cerevisiae* Y294[AG] and Y294[BBH1] strains, suggesting that there is scope for further improvements in enzyme production by *S. cerevisiae* Y294[AG]. However, when a more robust industrial amylolytic strain is constructed, additional exogenous enzyme might not be required. It should also be noted that Stargen 002 is a commercial enzyme preparation, which contains mainly amylases, but also other hydrolytic activities, such as cellulases and hemicellulases, not produced by *S. cerevisiae* Y294[AG]. The additional hydrolytic activities in Stargen 002 could assist in making the starches more available, but also release additional fermentable sugars from the cellulosic materials present in the PMS. This could explain the higher ethanol production by *S. cerevisiae* Y294[BBH] when Stargen 002 is added, compared to the ethanol produced by *S. cerevisiae* Y294[AG] (with no Stargen 002 addition) where the ethanol is solely derived from starch conversion.

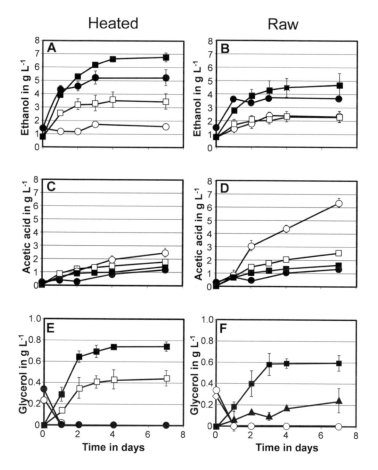

Figure 8.4: The *S. cerevisiae* Y294[BBH1] (—●— —○—) and Y295[AG] (—■— —□—) strains were cultivated on PMS as sole carbohydrate source under anaerobic conditions, without (open symbols), or with the addition of Stargen 002 enzymes (filled symbols). The (A, B) ethanol, (C, D) acetic acid, and (E, F) glycerol concentrations were monitored over time. Error bars represent standard deviations from triplicate experiments

CONCLUSIONS

The waste streams from industrial food processing facilities are potentially excellent carbohydrate resources. Apart from their use in the animal feed industry, they can also be used for the cultivation of microorganisms for the production of a variety of enzymes and commodities.

This preliminary study demonstrated (1) the use of potato waste and potato wastewater for the cultivation of fungal strains with the purpose of producing industrially relevant enzymes and (2) the use of an amylolytic *S. cerevisiae* strain for the production of bioethanol. The use of a bioreactor setup with controlled aeration would most likely improve the levels of enzyme activity obtained with the recombinant *A. niger* strains and ethanol production by the amylolytic *S. cerevisiae* strain when a higher solids loading is used. This study showed that a starch-rich waste can be utilised for the production of alternative bioproducts such as enzymes and bioethanol, highlighting the potential role that alternative biotechnological processes can add to the bioremediation and beneficiation of organic-rich waste streams. Waste streams that contain high quantities of fermentable carbohydrates should be identified for all industries in South Africa, listed in a national database and considered for implementation in an integrated biorefinery approach (Khan *et al.*, 2015). Where applicable, different waste streams can be combined, as demonstrated for laccase production. However, the transport of waste between the facilities should not negatively impact the overall cost of the process.

ACKNOWLEDGEMENTS

The authors would like to thank the Water Research Commission of South Africa for the financial support for Project K5/2225/3.

REFERENCES

Alriksson, B, Rose, SH, Van Zyl, WH, Sjöde, A, Nilvebrant, N-O and Jönsson, LJ. 2009. Cellulase production from spent lignocellulose hydrolysates with recombinant *Aspergillus niger. Applied and Environmental Microbiology,* 75, 2366–2374.

Bailey, MJ, Biely, P and Poutanen, K. 1992. Interlaboratory testing of methods for assay of xylanase activity. *Journal of Biotechnology,* 23, 257–270.

Biely, P, Mislovičová, D and Toman, R. 1985. Soluble chromogenic substrates for the assay of endo-1,4-β-xylanases and endo-1,4-β-glucanases. *Analytical Biochemistry,* 144, 142–146.

Bohlin, C, Jönsson LJ, Roth RL, and Van Zyl, WH. 2006. Heterologous expression of *Trametes versicolor* laccase in *Pichia pastoris* and *Aspergillus niger. Applied Biochemistry and Biotechnology,* 129–132, 195–214.

Catarino, J, Mendonça, E, Picado, A, Anselmo, A, Nobre da Costa, J, and Partidário, P. 2007. Getting value from wastewater: by-products recovery in a potato chips industry. *Journal of Cleaner Production,* 15, 927–931.

Cavka, AB, Alriksson, B, Rose, SH, Van Zyl, WH and Jönsson, LJ. 2011. Biorefining of wood: combined production of ethanol and xylanase from waste fiber sludge. *Journal of Industrial Microbiology and Biotechnology,* 38, 891–899.

Cavka, A, Alriksson, B, Rose, SH, Van Zyl, WH and Jönsson, LJ. 2014. Production of cellulosic ethanol and enzyme from waste fiber sludge using SSF, recycling of hydrolytic enzymes and yeast and recombinant cellulase-producing *Aspergillus niger. Journal of Industrial Microbiology and Biotechnology,* 41, 1191–1200.

Chimphango, AFA, Rose, SH, Van Zyl, WH and Görgens, JF. 2012. Production and characterisation of recombinant α-L-arabinofuranosidase for production of xylan hydrogels. *Applied Microbiology and Biotechnology,* 95, 101–112.

De Vries, RP and Visser, J. 2001. *Aspergillus* enzymes involved in degradation of plant cell wall polysaccharides. *Microbiology and Molecular Biology Reviews,* 65, 497–522.

García- García I, Venceslada JLB, Pena PRJ and Gómez ER. 1997. Biodegradation of phenol compounds in vinasse using *Aspergillus terreus* and *Geotrichum candidum. Water Research* 31, 2005–2011.

Hashem, M and Darwish, SMI. 2010. Production of bioethanol and associated by-products from potato starch residue stream by *Saccharomyces cerevisiae. Biomass Bioenergy,* 34, 953–959.

Jin, B, Van Leeuwen, HJ, Patel, B and Yu, Q. 1998. Utilisation of starch processing wastewater for production of microbial biomass protein and fungal α-amylase by *Aspergillus oryzae. Bioresource Technology,* 66, 201–206.

Kern, L, De Montigny, J, Jund, R and Lacroute, F. 1990. The *FUR1* gene of *Saccharomyces cerevisiae*: cloning, structure and expression of wild-type and mutant alleles. *Gene,* 88, 149–157.

Khan, N, Le Roes-Hill, M, Welz, PJ, Grandin, KA, Kudanga, T, Van Dyk, JS, Ohlhoff, C, Van Zyl, WH and Pletschke, BI. 2015. Fruit waste streams in South Africa and their potential role in developing a bio-economy. *South African Journal of Science,* 111 5/6, 1–11.

La Grange, DC, Pretorius, IS, Claeyssens, M and Van Zyl, WH. 2001. Degradation of xylan to D-xylose by recombinant *Saccharomyces cerevisiae* co-expressing the *Aspergillus niger* β-xylosidase *xlnD* and the *Trichoderma reesei* xylanase II *xyn2* genes. *Applied and Environmental Microbiology,* 67, 5512–5519.

Leighton, CS, Schönfeldt, HC, Visser, ER, Van Niekerk, J, Smith, MF and Morey, L. 2009. Mapping of different potato cultivars in South Africa. Available at: http://www.potatoes.co.za. [Accessed 8 October 2015].

Potato processing and uses. 2013. International Potato Center: Agricultural research for development. Available at: http://cipotato.org/potato/processing-uses. [Accessed 8 October 2015].

Potatoes South Africa. 2015. Representative of the potato producers. Available at: http://www.potatoes. co.za. [Accessed 8 October 2015].

DAFF (Department of Agriculture, Forestry and Fisheries) 2013. Production guideline.. Available at: http:// www.nda.agric.za/docs/Brochures/potatguidelines.pdf. [Accessed 8 October 2015].

Punt, PJ and Van den Hondel, CAMJJ. 1992. Transformation of filamentous fungi based on hygromycin B and phleomycin resistance markers. *Methods in Enzymology,* 216, 447–457.

Rose, SH and Van Zyl, WH. 2002. Constitutive expression of the *Trichoderma reesei* β-1,4-xylanase gene *xyn2* and the β-1,4-endoglucanase gene *egl* in *Aspergillus niger* in molasses and defined glucose media. *Applied Microbiology and Biotechnology,* 58, 461–468.

Rose, SH and Van Zyl, WH. 2008. Exploitation of *Aspergillus niger* for the heterologous production of cellulases and hemicellulases. *Open Biotechnology Journal,* 2, 167–175.

San Martin, D, Ramos, S and Jufia, J. 2016. Valorisation of food waste to produce new raw materials for animal feed. *Food Chemistry,* 198, 68–74.

Schieber, A, Stintzing, FC, and Carle, R. 2001. By-products of plant food processing as a source of functional compounds – recent developments. *Trends in Food Science and Technology,* 12, 401–413.

Strong PJ. 2011. Improved laccase production by *Trametes pubescens* MB89 in distillery wastewaters. *Enzyme Research* vol. 2011, Article ID 379176, doi:10.4061/2011/379176.

Strong, PJ and Burgess, JE. 2008. Fungal and enzymatic remediation of wine lees and five wine-related distillery wastewaters. *Bioresource Technology,* 99, 6134–6142.

Thomassen, LV, Larsen, DM, Mikkelsen, JD and Meyer, AS. 2011. Definition and characterization of enzymes for maximal biocatalytic solubilization of prebiotic polysaccharides from potato pulp. *Enzyme and Microbial Technology,* 49, 289–297.

Van Dyk, JS, Gama, R, Morrison, D, Swart, S and Pletschke, BI. 2013. Food processing waste: problems, current management and prospects for utilisation of the lignocellulose component through enzyme synergistic degradation. *Renewable and Sustainable Energy Reviews,* 26, 521–531.

Van Zyl, PJ, Moodley, V, Rose, SH, Roth, RL and Van Zyl WH. 2009. Production of the *Aspergillus aculeatus* endo-1,4-β-mannanase in *Aspergillus niger. Journal of Industrial Microbiology and Biotechnology,* 36, 611–617.

Viktor, M, Rose, SH, Van Zyl, WH and Viljoen-Bloom, M. 2013. Raw starch conversion by *Saccharomyces cerevisiae* expressing *Aspergillus tubingensis* amylase. *Biofuels for Biotechnology,* 6, 167.

MECHANICAL AND CHEMICAL TREATMENT

BENEFICIATION OF SAWDUST WASTE IN THE CONTEXT OF AN INTEGRATED FOREST BIOREFINERY MILL: KRAFT AND PRE-HYDROLYSIS KRAFT PULPING PROPERTIES

JE Andrew[1], J Johakimu[1], P Lekha[1], ME Gibril[1,2] and BB Sitholé[1,2]

[1] CSIR, Forestry and Forest Products Research Centre, PO Box 17001, Congella, Durban, 4013, South Africa

[2] University of KwaZulu-Natal, Discipline of Chemical Engineering, 4041, Durban, South Africa

Corresponding author e-mail: JAndrew@csir.co.za

ABSTRACT

As part of a broader objective to extract cellulose from sawdust waste material for the production of nanocrystalline cellulose, conventional industrially available processes such as the Kraft and pre-hydrolysis Kraft (PHK) processes were investigated for delignification of sawdust produced from *Eucalyptus grandis* wood. In the context of the integrated forest biorefinery, it was felt that it may be useful to provide South African papermakers with preliminary data on the Kraft and PHK pulping properties of sawdust as none appeared to be available in the country. The results showed that *E. grandis* sawdust Kraft pulps with acceptable yields (48%) and fibre morphologies comparable to conventional Kraft pulps, produced from woodchips, could be produced in the laboratory using typical Kraft pulping conditions. As expected, the exception was pulp strength properties such as burst, tear and tensile strengths, which were 50% to 70% lower than conventional pulps. During the pre-hydrolysis stage of the PHK process, up to 24 g.l^{-1} xylose could be removed from sawdust with minimal removal of lignin (0.1 g.l^{-1}) and cellulose (2.5 g.l^{-1}). Pulping of the pre-hydrolysed sawdust resulted in a pulp yield of ca. 35%. Preliminary characteristics measured on the unbleached PHK sawdust pulp such as pentosan content (3–4%), brightness (41%) and viscosity (760–850 ml.g^{-1}) alluded to its potential for the production of dissolving pulp.

Keywords: biorefinery, sawdust, Kraft pulp, pre-hydrolysis Kraft, nanocrystalline cellulose

INTRODUCTION

The forestry, timber, pulp, and paper (FTPP) sector plays an important role in the economy of South Africa and is a major contributor to job creation, directly employing over 170,000 people (Godsmark, 2014). The sector, however, like many others in the country, is facing severe challenges related to soaring energy costs and water shortages, some of which have a direct effect on production costs and international competitiveness. In addition, there is an ever-increasing pressure on the sector to make changes, improvements and/or adaptations to their processes to achieve cleaner production technologies that are more environmentally friendly. The disposal of waste in an economically and environmentally acceptable manner is another critical issue facing the sector. This is mainly due to increased difficulties in locating disposal facilities and complying with stringent environmental quality requirements imposed by waste management and disposal legislation.

A case in point is disposal of sawdust waste generated when wood is chipped, screened, sawed, turned, drilled or sanded. Sawmills are major producers of sawdust waste. Other typical industries that produce sawdust include pulp mills and furniture manufacturers. It is estimated that South Africa may have close to 400 sawmills of varying capacities countrywide (Timberwatch, 2000). Data on the amount of sawdust produced in the country is difficult to obtain, and where available, is severely outdated or vary significantly depending on the source. Based on figures obtained from Forestry South Africa (Godsmark, 2014), the total quantity of wood processed in the country during 2011/2012 was around 18.8 million tonnes, of which sawmills accounted for 20% or 3.76 million tonnes of this consumption. Olufemi *et al.,* (2012) estimated that only 56% of a log processed in a sawmill is recovered as sawn timber, while 44% is left as wood residues in the form of wood slab (34%) and sawdust (10%). Then, assuming that the amount of sawdust generated from sawn timber averages around 10%, it can be estimated that the amount of sawdust produced by South African sawmills was around 376,000 tonnes in 2012. These figures are conservative and exclude woodchips, offcuts, and shavings, and additional sawdust generated from pulp mills and other smaller informal sawmills. Estimates obtained from Timberwatch (2000) showed that at one stage, the forestry and wood processing industries in South Africa generated as much as 4 to 6 million tonnes of wood waste per year. According to one local sawmilling company, 90% of its revenue is obtained from only 40% of a tree (Herbst, 2013). This means that the remaining 60% of the tree may be discarded as waste products in the form of sawdust, chips, offcuts and shavings, or used to heat kilns to dry timber. Some waste products are also sold to composting companies and board manufacturers. However, there still appears to be an excess of sawdust that is either stockpiled on-site and allowed to biodegrade due to lack of disposal options or are sometimes landfilled. According to environmental regulations (e.g. Act 39 of 2004), these practices are being curtailed as they are environmental hazards that generate greenhouse gases (GHGs) and possible leaching of toxic chemicals into surrounding ground and water sources when stockpiled. In the case of landfilling as a means of waste disposal, significant costs are incurred by transporting waste to landfill sites, and maintaining and establishing new landfill sites.

Within the context of an integrated forest biorefinery, there is therefore a need to find new and innovative uses to beneficiate the sawdust waste. A biorefinery is a facility that integrates biomass conversion processes and equipment to produce fuels, power, heat and other value-added chemicals, in addition to traditional products (Van

Heiningen, 2006). The biorefinery concept is analogous to petroleum refineries that produce multiple fuels and speciality and commodity chemicals from petroleum. For the FTPP sector, producing "green" bioenergy, biochemicals and new biomaterials, in addition to traditional wood products, may therefore lead to competitive synergies, new markets and increased product flexibility for the sector, while at the same time mitigating some of its environmental impacts (Van Heiningen, 2006; Mäkinen, 2011). Within this context, the overall objective of the study was to investigate alternative options for beneficiation of sawdust into high value products. One option being explored by our research group is to produce nanocrystalline cellulose (NCC) from sawdust. Nanocrystalline cellulose is nano-sized cellulose particles that are extracted from cellulose. They have impressive mechanical properties that are comparable with stainless steel and Kevlar (George 2013), and thus have the potential to serve as the reinforcing (or load-bearing) components in composite materials. Due to their superior properties, they also offer promising opportunities for application in several areas, such as construction, automotive, medical and environmental (Shatkin *et al.*, 2014). However, because of the heterogeneity and low crystallinity of wood, NCC cannot be produced directly from it. A prerequisite is that the cellulose contained in the wood must first be fractionated and isolated from the remaining wood components such as lignin and the hemicelluloses (Abraham *et al.*, 2011). In this study, as part of the pre-treatment process to extract cellulose, the sawdust was delignified using a conventional industrial pulping process called the Kraft process. Although Kraft pulp from sawdust may be of poor quality with possibly low strength, it was considered that it might be useful to provide South African papermakers with preliminary data on the Kraft pulping properties of sawdust, as none appeared to be available in the country. Some possible uses of sawdust Kraft pulp include the production of tissue paper or addition to recycled paper pulps. With the inclusion of a preceding hydrolysis stage prior to Kraft pulping, i.e., the pre-hydrolysis Kraft (PHK) process, the process may also be used to produce dissolving pulps (Kautto *et al.*, 2010). Dissolving pulps are high-grade cellulose pulps with low amounts of hemicelluloses, lignin and degraded cellulose. They are used to manufacture several cellulose-derived products, such as cellulose acetate, viscose, microcrystalline cellulose and rayon (Sjöström, 1981). In addition, the pre-hydrolysate from the PHK pulping process is rich in hemicelluloses and may provide a valuable source of raw material for the production of additional products, such as ethanol, acetic acid and various polymers. The objective of this study was therefore to investigate the Kraft and pre-hydrolysis Kraft pulping properties of sawdust and to provide South African papermakers with preliminary data on the potential use of sawdust waste for Kraft and dissolving pulp applications.

BACKGROUND

During Kraft pulping, a uniform distribution of pulping chemicals throughout the woodchips is important for uniform delignification, which in turn produces pulps of high yield and quality (Zanuttini *et al.*, 2005). The penetration and diffusion of pulping chemicals throughout the woodchips is strongly affected by chip dimensions, such as length, width and thickness, with chip thickness being the principle dimension of concern (Jimenez *et al.*, 1990). Woodchips thicker than the ideal chip thickness of 2 mm to 8 mm can lead to increased rejects, while smaller chips can have a detrimental effect on pulp yield (MacLeod 2007). With a surface area 30 times that of conventional woodchips, the major challenge when pulping sawdust particles is the non-uniform wetting of these particles with cooking chemicals that lead to uneven penetration and diffusion of the pulping chemicals. In turn, this can lead to low pulp yield and strength properties (Winstead, 1972). To overcome this challenge, the conventional method of pulping sawdust involves pre-steaming, cooking and discharging of the sawdust pulp using specialised digesters of the Messing-Durkee (M&D) or Kamyr types (Luthe *et al.*, 2004; Korpinen *et al.*, 2006). When cooked in conventional woodchip digesters, sawdust is often added to woodchips in controlled amounts and co-cooked with the woodchips (Korpinen *et al.*, 2006). Although not practised in South Africa, sawdust pulping is common to many parts of the world, and has been practised since the early 1960s in the United States of America and Canada (Winstead, 1972; Korpinen & Fardim, 2009). Sawdust pulp yields can range from 43% to 45% for bleachable grades, and from 46% to 48% for unbleached pulps (Winstead, 1972). However, it has been found that sawdust pulp strength is significantly lower than that produced from woodchips (Winstead, 1972). This is presumably due to the shorter fibres obtained from the small sized sawdust particles. Despite this, sawdust pulps have been used in several papermaking applications. Korpinen & Fardim (2009) used sawdust Kraft pulp to reinforce thermomechanical (TMP) and pressurised ground-wood (PGW) pulps for use in uncoated super-calendered (SC), and lightweight coated (LWC) papers. They found that up to 30% of sawdust Kraft pulp could be added to TMP and PGW pulps, without negatively impacting any of its properties. Winstead (1972) also reported that up to 20% sawdust pulp could be added to fine papers without any detrimental effect on strength properties; and that 50% to 70% of sawdust pulp may be blended to produce towelling and tissue papers.

With the addition of a preceding hydrolysis stage prior to Kraft pulping, the Kraft process can also be used to produce dissolving pulp i.e., the pre-hydrolysis Kraft (PHK) process (Kautto *et al.*, 2010). Pre-hydrolysis of woodchips with dilute acids (Saukkonen *et al.*, 2012), bases (Um & Van Walsum, 2009; Johakimu & Andrew, 2013), hot water (Sixta *et al.*, 2013) and steam (Leschinsky *et al.*, 2009) enable the extraction and recovery of the hemicelluloses traditionally burned with lignin in the black liquor.

EXPERIMENTAL

Sampling and sample preparation

One-and-half metre bottom billets were taken from each of five 11-year-old *Eucalyptus grandis* trees sampled from a single site (site index 24). The billets were chipped and sawdust was generated by passing the chips, first through a hammer mill, and then through a Wiley mill. The sawdust was collected in a plastic bag and stored at room temperature until required.

Kraft pulping

Pulping was carried out in an electrically heated rotating digester (Regmed, Brazil) using the Kraft process. The digester consisted of a 25 L stainless steel main boiler vessel that housed four intermediate 1.5 L boiler vessels, which were independent of the main boiler vessel and from each other. 150 g of oven dried (OD) equivalent sawdust samples were pulped at 170°C for 20, 40, 50, and 60 minutes. The ramp time to maximum temperature was 90 minutes. The sulphidity of the cooking liquor was kept constant at 27%, and the active alkali charge (%AA as Na_2O) was varied at three levels (14%, 16% and 18%). During pulping, the moisture content of the sawdust was taken into account in order to maintain a constant liquor-to-wood (L:W) ratio, which was varied at two levels (4.5:1 and 7.5:1). After pulping, the cooked sawdust pulp was disintegrated for five minutes using a standard laboratory pulp disintegrator before washing under vacuum on a Büchner flask and funnel using tap water. The pulp was then screened on a 0.15 mm slotted screen to separate uncooked sawdust material (rejects) from the pulp.

Pre-hydrolysis Kraft (PHK) pulping

Water was used for the pre-hydrolysis treatment. The conditions employed entailed a temperature of 170°C, with a ramp time of 60 minutes to maximum temperature. The time at maximum temperature was varied between 15 to 60 minutes. After pre-treatment, the sawdust material was removed from the reactor and filtered and washed under vacuum on a Büchner flask with deionised water. Prior to washing, an aliquot of the pre-hydrolysate was taken for chemical analysis. After washing, the pre-extracted sawdust was weighed and the moisture content measured to calculate the yield and to maintain an accurate L:W ratio during the subsequent pulping. Fresh white liquor was then added to the pre-hydrolysed sawdust and conventional Kraft pulping was carried out as described above.

Chemical characterisation

The polysaccharide content of the wood or pulp was determined by acid hydrolysis (TAPPI T249 cm-85) followed by separation using high-performance anion exchange chromatography, coupled with pulsed ampherometric detection (Wright & Wallis 1996; Wallis *et al*., 1996). Acid insoluble lignin was determined by quantitatively filtering the hydrolysate from the acid hydrolysis step through a 0.45 µm filter paper. The material remaining on the filter paper was defined as Klason lignin. The acid soluble lignin was determined by measuring the UV absorbance of the filtrate or decantate at 205 nm using a Cary spectrophotometer (Varian, USA). The acid soluble lignin and monosaccharides in the pre-hydrolysate was measured by first acidifying the extract with 1 ml of 72% sulphuric acid followed by hydrolysis in an autoclave with 4% H_2SO_4 at 121 °C (103 kPa) for 1 hour prior to analysis. Cellulose was determined according to the Seifert method (Fengel & Wegener, 1984). Ash content and Kappa number were measured using TAPPI standard methods T211 om-93 and T236 cm-85, respectively. The intrinsic viscosity of pulps was measured using the Scandinavian method SCAN-CM 15:88

(1988). For the PHK pulps, the pentosan content was measured using a method based on TAPPI T450 os-44.

Fibre morphology and paper properties

Fibre morphological properties were measured using a Morfi Compact fibre analyser (Techpap, France). Pulp properties were assessed by preparing and testing handsheets. Handsheets with a basis weight of 60 g.m^{-2} were prepared on a Rapid Köthen sheet forming machine. Pre-conditioned handsheets (23°C, 50% RH) were tested for burst (TAPPI T403 om-02), tear (T414 om-98), tensile (T494 om-01), tensile energy absorption (T494 om-01), breaking length (T494 om-01), sheet density (T411 om-97), stiffness (T489 om-08), porosity (ISO 5636-3), water absorptiveness (T441 om-98) and brightness (T452 om-02).

RESULTS AND DISCUSSION

Sawdust Kraft pulping properties

Screened pulp yield (SPY) generally increased with increasing cooking times (Figure 9.1A). Using a constant AA charge (i.e., varying L:W ratio), optimum SPY was obtained using 7.5:1 L:W ratio at 60 minutes cooking time. This was expected as a higher liquid volume ensured a more efficient wetting of the material. Subsequently, this ensured uniform distribution of the cooking liquor and as a result, a more uniform pulp (Winstead, 1972). At constant cooking time of 60 minutes and 7.5:1 L:W ratio, the SPY increased with increasing AA. This was consistent with findings by MacLeod & Kingsland (1990). Optimum SPY (48.4%) was achieved using 60 minutes cooking time, 7.5:1 L:W ratio and 18% AA. This yield fell within the typical range (45-55%) for bleachable grade hardwood Kraft pulps (Macleod, 2007), and was only a few percentage points below that obtained for other South African *E. grandis* Kraft pulps produced from woodchips (Sefara *et al.*, 2000; Megown *et al.*, 2000; Andrew *et al.*, 2014).

The reject content (Figure 9.1B) decreased sharply as cooking time increased. At shorter cooking times, below 40 minutes, the sawdust material pulped using 4.5:1 L:W ratio showed lower reject contents compared to that pulped using the 7.5:1 L:W ratio. Presumably, at the shorter pulping times, the lower reject content may be attributed to the higher concentration of pulping chemicals in contact with the sawdust material when a lower L:W ratio is used. As the pulping time was increased when using 7.5:1 L:W ratio, the more uniform pulping achieved was related to a better wetting of the sawdust material due to a higher amount of free liquid available. At a constant cooking time of 60 minutes and 7.5:1 L:W ratio, as expected, the reject content increased as the AA decreased. This may be due to insufficient chemicals available for delignification.

The Kappa number is a measure of the residual lignin in the pulp and is usually used to estimate the amount of chemicals required during bleaching. For hardwood bleachable grade pulps, as in the case of Eucalypt pulps, the Kappa number target during pulping is typically around 16 to 18 but can be as low as 12 in some instances.

For the sawdust pulps, the Kappa number decreased as cooking time increased (Figure 9.1C). Lower Kappa numbers were achieved when pulping at the higher L:W ratio of 7.5:1. This may be an indication of more uniform cooking at the higher L:W ratio. At a constant cooking time of 60 minutes and 7.5:1 L:W ratio, delignification was reduced as AA dosage decreased, as shown by the higher Kappa numbers obtained.

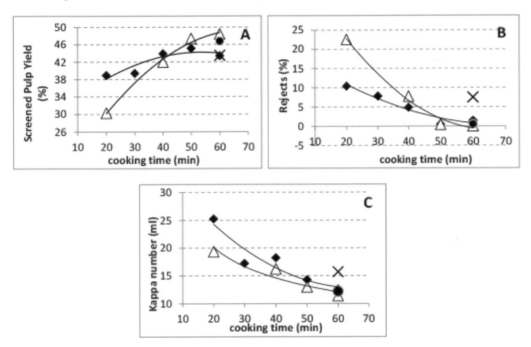

Figure 9.1: Kraft pulping properties of sawdust waste material: 4.5L:W-18%AA (♦); 7.5L:W-18% AA (Δ); 7.5L:W-16%AA (●) and 7.5L:W-14%AA (x)

Sawdust Kraft pulp fibre morphology

The fibre content of the sawdust pulps averaged 26.8 mil fibres.g^{-1} and appeared to be in a similar range as *Eucalyptus* Kraft pulps produced from woodchips (Table 9.1). Neiva *et al.,* (2015) reported 24–26 mil fibres.g^{-1} for *E. grandis, E. saligna, E. globulus, E. propinqua* and *E. botryoides* pulps. However, they found that the fibre content of Eucalypt pulps produced from chips could vary significantly. They reported values as low as 18 mil fibres.g^{-1} in the case of *E. maculata* and as high as 41 mil fibres.g^{-1} for *E. viminalis.* Fibre length is an important property because a minimum length is required for inter-fibre bonding during papermaking (Paavilainen, 1993). It is generally accepted that pulp strength is strongly correlated to fibre length. The sawdust pulp fibres exhibited an average length-weighted fibre length of ca. 0.65 mm, which is within the acceptable range (0.6–0.85 mm) for Eucalypt pulps produced from woodchips (Foelkel, 2007) and higher than that reported for *E. camaldulensis* (0.57 mm), *E. sideroxylan* (0.57 mm), *E. viminalis* (0.6 mm), *E. ovata* (0.61 mm), *E. rudis* (0.63 mm), and *E. resinifera* (0.63 mm) pulps produced from woodchips (Neiva *et al.*, 2015). Only a few Eucalypt species such as *E. saligna, E. botryoides, E. globulus, E. maculata* and *E. grandis* were reported with higher length-weighted pulp fibre lengths of 0.71 mm, 0.72 mm, 0.73 mm, 0.75 mm and 0.76 mm, respectively (Neiva *et al.*, 2015). Fibre coarseness is an important papermaking property and is strongly related to sheet structure and formation, which in turn, affects the strength properties, sheet density and porosity of paper (Ramezani & Nazhad, 2004). A higher coarseness value is indicative of fibres with thicker cell walls that are stiffer and less collapsible (Ramezani & Nazhad, 2004). Sawdust pulp fibres exhibited fibre coarseness values that were in a similar range as conventional *Eucalyptus* Kraft pulp fibres produced from woodchips. Neiva *et al.* (2015) reported typical values for coarseness for several *Eucalyptus* Kraft pulp species that ranged between 4.6 and 8.4 mg.100m^{-1}. They reported an overall average coarseness value of 6.4 mg.100m^{-1} for these Eucalypts, and that of *E. grandis* pulps in particular was 6.1 mg.100m^{-1}. Similar to findings of other researchers, the coarseness of sawdust Kraft pulps increased with increasing fibre length, and as expected, it correlated with the fibre content of the sawdust pulps, in that the number of fibres per gram of pulp increased with decreasing fibre coarseness. Fibre deformations, such as kinks and curls may contribute to reduced fibre strength. According to Foelkel (2007), a typical range for conventional Eucalyptus Kraft pulp kinks is between 0.4 and 1.5 kinks, with kink angles greater than 30°. For fibre curliness, the range given by Foelkel (2007) was between 5 and 15%.

Table 9.1: Sawdust Kraft pulp fibre morphology

Fibre characteristics	4.5L:W-18%AA	7.5L:W-18%AA	7.5L:W-16%AA	7.5L:W-14%AA	Conventional pulps
Number of fibres	5106	5113	5099	5086	-
Fibre content (x10^6.g^{-1})	27.3	30.0	25.0	24.7	24–26[a]
Length-weighted length (µm)	644	644	654	676	600–850[b]
Fibre coarseness (mg.100m^{-1})	5.9	6.4	6.9	7.0	4.6–8.4[a]
Kink number	1.2	1.3	1.2	1.2	0.4–1.5[b]
Kink angle (°)	132.3	130.7	132.0	131.0	>30.0[b]
Fibre curl (%)	5.0	6.6	5.4	5.7	5–15%[b]

[a] *Neiva et al., 2015*
[b] *Foelkel 2007*

Sawdust Kraft pulp and paper properties

The sawdust Kraft pulp handsheet properties at constant freeness of 400 ml are listed in Table 9.2. Where available, they were compared to conventional South African *E. grandis* Kraft pulps produced from woodchips. While an extensive set of paper properties is provided for information purposes, only a few pertinent properties related to paper strength are discussed. Sheet density is an important property that affects most mechanical and physical properties of paper. The density of the handsheets produced from sawdust Kraft pulps was about 35% lower than the average sheet densities listed for conventional Kraft pulps in Table 9.2. Generally, a higher sheet density is indicative of better bonding in a paper sheet. Sheet density usually increases with beating due to improved flexibility of the beaten fibres that conform easily to one another in the sheet. This results in an increased bonded area and more densely packed fibres in the paper sheet (Law, 1999; Gulsoy *et al.*, 2013). To reach 400 ml CSF, as reported in Table 9.2, the sawdust Kraft pulp required minimal beating (650 beating revolutions). This was 60% lower than that required by conventional Kraft pulps produced from woodchips. The increased degree of beating of conventional pulps therefore resulted in a more compact sheet and higher sheet density for the conventional Kraft pulps.

Directly related to this is tensile strength, which in addition to being dependant on the bonding area (O'Neil *et al.*, 1999), is also dependant on fibre length (O'Neill *et al.*, 1999), fibre strength (Gulsoy *et al.*, 2013) and bonding strength (Twimasi *et al.*, 1996). The tensile strength of the handsheets produced from sawdust pulps were found to be ca. 50% lower than handsheets produced from conventional Kraft pulps. Similar findings were observed for bursting and tearing strengths, which were 70% and 50% lower than conventional Kraft pulps, respectively. Lower bursting strength is attributed to weaker fibre-to-fibre bonding, and as a result, a more bulky sheet (Smook, 1992); while a decrease in tearing strength is related to a lower fibre strength and shorter fibres (Muneri, 1994).

Chemical characterisation of the pre-hydrolysate during PHK pulping

In hardwoods, the hemicelluloses account for 25% to 35% of the wood. However, during pulping, a majority of the hemicelluloses are removed from the wood and collected in the black liquor where it is burned together with lignin. In recent times, there has been increasing interest in the valorisation of these sugars since their contribution to the calorific value of the black liquor is not significant (Van Heiningen, 2006). Within the context of resource optimisation and a biorefinery mill, the attractiveness of the PHK process is due to the availability of the hemicellulose-rich stream for beneficiation into a range of additional products (Kautto *et al.*, 2010). Table 9.3 shows the chemical composition of the pre-hydrolysate extract prior to pulping. Maximum arabinoses and rhamnoses were removed early during pre-hydrolysis, and their concentrations in the pre-hydrolysate remained relatively constant thereafter over the entire duration of the pre-hydrolysis stage. The amount of the galactoses removed also appeared relatively stable early in the pre-hydrolysis stage between 15 to 45 minutes, and then increased marginally around 60 minutes.

In the case of glucose, ca. 2 $g.l^{-1}$ was removed early in the pre-hydrolysis stage, and this increased marginally as the pre-hydrolysis time increased, peaking at ca. 2.5 $g.l^{-1}$ after 60 minutes in the pre-hydrolysate. Xylose was the dominant monosaccharide removed during pre-hydrolysis. At 15 minutes pre-hydrolysis time, up to 17 $g.l^{-1}$ was found in the pre-hydrolysate, and this increased to just over 24 $g.l^{-1}$ after 60 minutes pre-hydrolysis time. The removal of lignin was also marginal, with the amount removed remaining fairly constant at ca. 0.1 $g.l^{-1}$ over the entire duration of the pre-hydrolysis stage. Similar to other studies (Tunc *et al.*, 2008), the pH of the extract varied between 2 and 3. The high acidity of the extract was attributed to the hydrolysis of the acetyl groups attached to the hemicelluloses that result in the formation of acetic acid, hence lowering the pH of the pre-hydrolysate extract (Pu *et al.*, 2011).

Table 9.2: Handsheet properties of sawdust Kraft pulps compared at constant freeness of 400 ml CSF to Kraft pulps produced from woodchips

	Sawdust pulp	Woodchip pulp
Beatings to reach 400 ml CSF (revs)	650	1750[a]
Sheet density (kg.m^{-3})	498 (±10)	792[a] 76[b] 740[c]
Tensile index (kNm.kg^{-1})	51.5 (±2.3)	92[a] 102[b] 105[c]
Burst index (MN.kg^{-1})	2.1 (±0.3)	7.3[b] 7.0[c]
Tear index (Nm^2.kg^{-1})	4.7 (±0.1)	9.6[a] 8.6[b] 9.0
Tensile energy of adsorption (J.m^{-2})	38.7 (±2.1)	na
Water absorptiveness at 45 secs (g.m^{-2})	132 (±8)	na
Stiffness (mN)	737 (±65)	na
Porosity (ml.min^{-1})	1585 (±129)	1250[d]
Brightness (%ISO)	40.3 (±0.7)	na
Unbeaten pulp brightness, (%ISO)	42.4 (±0.7)	na

na – not available
[a] Azeez, Andrew & Sithole, 2015
[b] Megown *et al.* (2000)
[c] Grzeskowiak *et al.* (2000)
[d] Neiva *et al.* (2015) and reported at ~550 ml CSF

Table 9.3: Chemical characterisation of the pre-hydrolysate liquors prior to PHK pulping

Hydrolysis time at 170°C (min)	pH	Arab (mg.l^{-1})	Gal (mg.l^{-1})	Rham (mg.l^{-1})	Glu (mg.l^{-1})	Xyl (mg.l^{-1})	Man (mg.l^{-1})	Acid soluble lignin (mg.l^{-1})
15	3.08	503	1406	665	1828	17511	608	98
30	3.11	405	1497	756	2309	20465	512	103
45	2.39	237	1471	748	2357	21936	710	106
60	1.94	451	1714	813	2476	23453	463	110

Chemical characterisation of the pre-hydrolysed sawdust during PHK pulping

The sawdust yield remaining after pre-hydrolysis decreased with increasing pre-hydrolysis time (Table 9.4). At 15 minutes, the yield loss was already ca. 20%. After 60 minutes, this loss extended to 28%. Similar results were obtained by other researchers when hydrolysing woodchips at 170°C (Pu *et al.*, 2011). The measured cellulose content in the pre-hydrolysed sawdust remained stable during pre-hydrolysis and averaged ca. 62%. More than 50% of the original xyloses in the wood were lost early in the pre-hydrolysis stage (after 15 minutes). Beyond this time, the xylose concentration steadily decreased to ca. 2.5% after 60 minutes. Maximum removal of the arabinoses and galactoses occurred early in the pre-hydrolysis stage, and their residual concentrations were negligible in the pre-extracted wood. Beyond 30 minutes pre-hydrolysis time, virtually all the rhamnoses were removed and none were detected in the pre-extracted wood after 45 minutes. Mannose appeared to be more resistant to pre-hydrolysis, and after 60 minutes, 0.3% remained in the pre-extracted wood. Interestingly, not all the soluble lignin was removed, and around 3% remained in the wood after pre-hydrolysis. The Klason lignin fell 1% to 35% after 60 minutes. The increase in glucose in the pre-extracted sawdust was due to the removal of the other sugars and some lignin. After 60 minutes pre-hydrolysis, the glucose concentration in the pre-hydrolysed wood reached 57%.

Expressed as a percentage of the original wood, Figure 9.2 shows that the rhamnoses were removed completely from the wood during the pre-hydrolysis stage (i.e. 100% removal). This was followed closely by the galactoses (97% removal), arabinoses (88%), xyloses (80%), and to a lesser extent, the mannoses (36%), and finally the glucose, with around 12% removed. Besides the glucoses, the two other main monosaccharides remaining in the wood after pre-extraction were the xyloses, at 1.8% residual concentration, and the mannoses at 0.2%.

Kraft pulping of pre-hydrolysed sawdust

The optimum Kraft pulping conditions obtained for pulping the un-hydrolysed sawdust (i.e., conventional Kraft pulping process) was 18% AA, 7.5:1 LW and 60 minutes pulping time. These optimum conditions were then used to pulp the pre-hydrolysed sawdust material. The resulting Kraft pulping properties of the pre-hydrolysed sawdust material is shown in Table 9.5. The screened pulp yield varied by about 1% around 42% to 43% for all pre-hydrolysis treatments. As a percentage of the original sawdust material, this equated to an overall Kraft pulp yield of ca. 35% for the sawdust pre-hydrolysed between 15 to 45 minutes. Extending the pre-hydrolysis beyond these times resulted in a further 2% drop in pulp yield to 33%. No uncooked sawdust material was detected after screening. It can be seen from the lower Kappa numbers that the rate of delignification was significantly higher for the pre-hydrolysed sawdust compared to conventional Kraft pulping. Similar results have been reported previously, in that lignin removal is enhanced during Kraft pulping of pre-hydrolysed wood (Kautto *et al.*, 2010; Duarte *et al.*, 2011; Johakimu & Andrew, 2013).

Table 9.4: Chemical characterisation of pre-hydrolysed sawdust prior to PHK pulping.

Hydrolysis time at 170°C (min)	Sawdust yield (%)	Cell (%)	Arab (%)	Gal (%)	Rham (%)	Glu (%)	Xyl (%)	Man (%)	Acid soluble lignin (%)	Klason lignin (%)
15	80.4	61.4	0.02	0.15	0.07	54.0	4.42	0.45	3.70	36.5
30	75.0	62.4	0.01	0.06	0.04	55.6	3.01	0.37	3.31	37.0
45	73.0	62.8	0.02	0.02	0.00	56.0	2.99	0.34	3.13	36.9
60	72.1	62.2	0.02	0.03	0.00	56.9	2.49	0.30	3.13	35.4

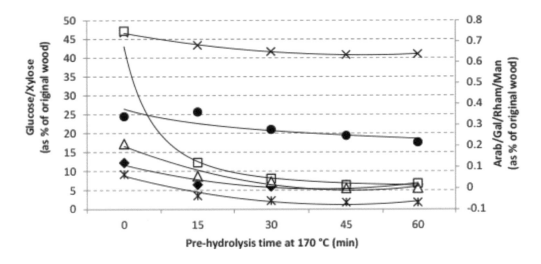

Figure 9.2: Residual hemicelluloses content in pre-hydrolysed sawdust material prior to PHK pulping: glucose (×), arabinose (♦), mannose (●), galactose (□), xylose (ж) and rhamnose (Δ)

Pentosans in dissolving pulps are considered contaminants for the production of cellulose derivatives. Their residual concentration in the unbleached PHK sawdust pulps ranged between 3% and 4%, and showed a decreasing trend as the severity of the pre-hydrolysis stage was increased. A similar trend was observed for viscosity, which decreased from 848 ml.g^{-1} to 759 ml.g^{-1}, when the pre-hydrolysis time was increased from 15 to 60 minutes. Brightness, on the other hand, increased with increasing severity of the pre-hydrolysis stage. Maximum brightness of 41.4% was achieved after 60 minutes pre-hydrolysis time. Sixta and Schild (2009) reported similar results for pulp yield, Kappa number, brightness and pentosan content of unbleached PHK pulps produced from *E. globulus* woodchips. However, the viscosity they reported was about 25% higher than that reported in this study for sawdust PHK pulps.

CONCLUSION AND WAY FORWARD

As part of a broader objective to isolate cellulose from sawdust waste material for the production of nanocrystalline cellulose, conventional industrially available processes such as the Kraft and pre-hydrolysis Kraft (PHK) processes were investigated for delignification of sawdust produced from *E. grandis* wood. Sawdust Kraft pulp yields (48%) and fibre morphologies were within acceptable ranges and comparable to conventional Kraft pulps produced from woodchips. The exception was the pulp strength properties such as burst, tear and tensile strengths, which were 50% to 70% lower than conventional pulps. Possible applications of this pulp could be addition or replacement of thermomechanical pulps and/or chemical pulps during newsprint manufacture, or other recycled paper applications such as tissue paper.

Table 9.5: Kraft pulping properties of pre-hydrolysed sawdust

Pre-hydrolysis time at 170°C (min)	Screen pulp yield (%)	Screen pulp yield (as% of original sawdust)	Reject content (%)	Total pulp yield (%)	Kappa number (ml)	Viscosity (ml/g)	Pentosan content (%)	Brightness (ISO)
15	42.5	35.2	0.0	42.5	5.3	848	3.77	33.5
30	42.8	35.3	0.0	42.8	3.1	791	3.50	34.0
45	43.3	35.2	0.0	43.3	3.0	759	3.45	37.5
60	42.3	32.7	0.0	42.3	2.6	759	3.08	41.4

During the pre-hydrolysis stage of the PHK process, up to 24 g.l^{-1} xylose could be removed from sawdust with minimal removal of lignin (0.1 g.l^{-1}) and cellulose (2.5 g.l^{-1}). Wood yield after the pre-hydrolysis stage ranged between 70% and 80%, and pulping of the pre-hydrolysed sawdust resulted in pulp yield ca. 35%. Preliminary properties measured on the unbleached PHK sawdust pulps such as pentosan content (3–4%), brightness (41%) and viscosity (760-850 ml.g^{-1}) alluded to its potential for the production of dissolving pulps. Future work in this area may include bleaching and full characterisation of pulps, and may be extended to include techno-economic studies for beneficiation of the hemicelluloses, lignin and cellulose fractionated from sawdust. Co-cooking of sawdust with woodchips is another option to explore.

ACKNOWLEDGEMENTS

A special thank you to SAPPI for assistance with measurement of pentosan content in PHK pulps. Thanks and appreciation to CSIR technical staff for assistance with the remaining analyses.

REFERENCES

Abraham, E, Deepa, B, Pothan, LA, Jacob, M, Thomas, S, Cvelbar, U and Anandjiwala, R. 2011. Extraction of nanocellulose fibrils from loignocellulosic fibres: a novel approach. *Carbohydrate Polymers,* 86, 1468–1475.

Andrew, JE, Johakimu, J and Ngema, NE. 2013. Ozone bleaching of South African *Eucalyptus grandis* Kraft pulps containing high levels of hexenuronic acids. *TAPPI Journal,* 12(8), 7–14.

Andrew, JE, Johakimu, J and Sithole, BB. 2014. Bleaching of Kraft pulps produced from green liquor pre-hydrolyzed South African *Eucalyptus grandis* wood chips. *Nordic Pulp & Paper Research Journal,* 29(3), 383–391.

Azeez, MA, Andrew, JE and Sithole, BB. 2016. A preliminary investigation of Nigerian Gmelina arborea and Bambusa vulgaris for pulp and paper production. *Maderas-Ciencia Tecnología,* 18(1), 65–78.

Duarte, GV, Ramarao, BV, Amidon, TE and Ferreira, PT. 2011. Effects of hot water extraction on hardwood Kraft pulp fibres (*Acer saccharum,* Sugar Maple). *Ind. Eng. Chem. Res.* 50, 9949–9959.

Fengel, D, and Wegener, G. 1984. Wood: Chemistry, ultrastructure, reactions. Walter de Gruyter & Co., Berlin pp. 6–25.

Foelkel, C. 2007. The Eucalyptus fibers and the Kraft pulp quality requirements for paper manufacturing. Eucalyptus Online Book and Newsletter. Available at: http://www.eucalyptus.com.br/capitulos/ENG03_fibers.pdf [Accessed 23 September 2015].

George, A. 2013. New wood pulp concoction stronger than Kevlar, carbon fiber. Available at: http://www.wired.com/2012/09/wood-pulp-material. [Accessed 12 July 2016].

Godsmark, R. 2014. The South African industry's perspective on Forestry & forest products statistics – 2014. Available at: http://www.forestry.co.za/statistical-data [Accessed 5 October 2015].

Gulsoy, SK, Kustas, S and Erenturk, S. 2013. The effect of old corrugated container (OCC) pulp addition on the properties of paper made with virgin softwood Kraft pulps. *Bioresources,* 8(4), 5842–5849.

Grzeskowiak, V, Turner, P and Megown, RA. 2000. The use of densitometry and image analysis techniques to predict pulp strength properties in Eucalyptus plantations. *TAPPSA Conference "African Paper Week 2000 and Beyond", Durban, South Africa, 17–20 October.*

Herbst, S. 2013. Forestry group moves ahead plan to integrate Sabie site, extract more value. Available at: http://www.engineeringnews.co.za/article/forestry-group-moves-ahead-plan-integrate-sabie-site-extract-more-value-2013-05-10 [Accessed 5 October 2015].

Jimenez, G, Chian, DS, McKean, WT and Gustafson, R.R. 1990. Experimental and theoretical studies to improve pulp uniformity. *Proceeding of the 1990 TAPPI Pulping Conference, Toronto, Canada,* 49–53.

Johakimu, J and Andrew, JE. 2013. Hemicellulose extraction from South African *eucalyptus grandis* using green liquor and its impact on Kraft pulp efficiency and paper making properties. *Bioresources,* 8(3), 3490–3504.

Kautto, J, Saukkonen, E and Henricson, K. 2010. Digestability and paper-making properties of prehydrolyzed softwood chips. *Bioresources,* 5(4),2502–2519.

Korpinen, R, Hultholm, T, Lönnberg, B and Achrén, S. 2006. Development of sawdust cooking. *Appita Journal,* 59(5), 406–411.

Korpinen, RI and Fardim, PE. 2009. Reinforcement potential of bleached sawdust Kraft pulp in different mechanical pulp furnishes. *Bioresources,* 4(4), 1572–1585.

Leschinsky, M, Sixta, H and Patt, R. 2009. Detailed mass balances of the autohydrolysis of *Eucalyptus globulus* at 170°C. *Bioresources,* 4(2), 687–703.

Luthe, C, Berry, R and Li, J. 2004. Polysulphide for yield enhancement in sawdust pulping: Does it work? *Pulp and Paper Canada,* 105(1), 32–37.

MacLeod, JM and Kingsland, KA. 1990. Kraft-AQ pulping of sawdust. *TAPPI Journal,* 73(1), 191–193.

MacLeod, M. 2007. The top ten factors in Kraft pulp yield. *Paperi ja Puu – Paper and Timber,* 89(4), 1–7.

Mäkinen, T. 2011. Biorefineries in pulp and paper industry. *Central European Biomass Conference, 26–29 January, Graz, Austria.*

Megown, KA, Turner, P, Male JR and Retief, J. 2000. The impact of site index and age on the wood, pulp and pulping properties of a Eucalyptus grandis clone (Tag 5). *IUFRO Conference, 8–13 October, Durban, South Africa.*

Neiva, DM, Fernandes, L, Araújo, S, Lourenço, A, Gominho, J, Simões, R and Pereira H. 2015. Pulping potential of young eucalypts: a comparative study of wood and pulp properties of 12 eucalypt species. *7th International Colloquium on Eucalyptus Pulp, 26–29 May, Vitória, Espirito Santo, Brazil.*

Olufemi, B, Akindeni, JO and Olaniran, SO. 2012. Lumber recovery efficiency among selected sawmills in Akure, Nigeria. *Drvna Industrija,* 63(1), 15–18.

Paavilainen, L. 1993. Importance of cross-dimensional fibre properties and coarseness for characterization of softwood sulphate pulp. *Paperi ja Puu,* 75(5), 343–351.

Pu, Y, Treasure, T, Gonzalez, R, Venditti, R and Jameel, H. 2011. Autohydrolysis pretreatment of mixed hardwoods to extract value prior to combustion. *Bioresources,* 6(4), 4856–4870.

Ramezani, O and Nazhad, MM. 2004. The effect of coarseness on paper formation. *African Pulp and Paper Week, 12–15 October, Durban, South Africa.*

Saukkonen, E, Kautto, J, Irina, R and Backfolk, K. 2012. Characteristics of prehydrolysis-kraft pulp fibers from Scots pine. *Holzforschung,* 66(7), 801–808.

Sefara, NL, Conradie, D and Turner, P. 2000. Progress in the use of near-infrared absorption spectroscopy as a tool for the rapid determination of pulp yield in plantation eucalypts. *TAPPSA Journal,* November, 15–17.

Shatkin, JA, Wegner, TH, Bilek, EM and Cowie, J. 2014. Market projections of cellulose nanomaterial-enabled products – Part 1: Applications. *TAPPI Journal,* 13(5), 9–16.

Sixta, H, Iakovlev, M, Testova, L, Roselli, A, Hummel, M, Borrega, M, Van Heinigen, A, Froschauer, C and Schottenberger, H. 2013. Novel concepts of dissolving pulp production. *Cellulose,* 20(4), 1547–1561.

Sixta, H and Schild G. 2009. A new generation Kraft process. *Lenzinger Berichte,* 87, 26–37.

Sjöström, E. 1981. Wood chemistry: fundamentals and applications, Academic Press, New York, 169–189.

Smook, GA. 2002. Handbank for pulp and paper technologists. Angus Wilde Publications, Vancouver.

Timberwatch 2000. The state of forestry in South Africa today. Available at: http://www.timberwatch.org/old_site/archives/2000807stateofforrestry.htm [Accessed 15 October 2015].

Tunc, MS and Heiningen, ARP. 2008. Hemicellulose extraction of mixed southern hardwood with water at 150°C: effect of time. *Industrial & Engineering Chemistry Research,* 47, 7031–7037.

Um, B-H and Van Walsum, GP. 2009. Acid hydrolysis of hemicellulose in green liquor pre-pulping extract of mixed northern hardwoods. *Applied Biochemistry and Biotechnology,* 153, 127–138.

Van Heiningen, A 2006. Converting a Kraft pulp mill into an integrated forest biorefinery, *Pulp & Paper Canada,* 107(6), 38–43.

Wallis, AFA, Wearne, RH and Wright, PJ. 1996. Chemical analysis of polysaccharides in plantation eucalypt woods and pulps. *Appita Journal,* 49(4), 258–262.

Winstead, TE. 1972. Rapid Kraft pulping of sawdust and other small wood particles. *Paper Trade Journal,* 156(43), 52–53.

Wright, PJ and Wallis, AFA. 1996. Rapid determination of carbohydrates in hardwoods. *Holzforschung,* 50, 518–524.

Zanuttini, M, Citroni, M, Marzocchi, V and Inalbon, C. 2005. Alkali impregnation of hardwood chips. *TAPPI Journal,* 4(2), 28–30.

DEVELOPMENT OF SUSTAINABLE BIOBASED POLYMER AND BIO-NANOCOMPOSITE MATERIALS USING NANOCELLULOSE OBTAINED FROM AGRICULTURAL BIOMASS

A Mtibe,[1] S Muniyasamy[1,2] and TE Motaung[3]

[1] CSIR Materials Science and Manufacturing, Polymers and Composites, P.O. Box 1124, Port Elizabeth, 6000, South Africa

[2] Department of Textile Science, Faculty of Science, Nelson Mandela University, P.O. Box 77000, Port Elizabeth, 6031, South Africa

[3] Department of Chemistry, Faculty of Science and Agriculture, University of Zululand, Private Bag X1001, KwaDlangezwa, 3886, South Africa

Corresponding author e-mail: SMuniyasamy@csir.co.za

ABSTRACT

Biobased polymer and bio-nanocomposites have provided significant improvement in material science, moving towards the development of green materials to replace petro-based materials. The present study investigated the value-added utilisation of agricultural biomass residues derived from sugar cane bagasse and maize stalks for the development of biobased polymer and bio-nanocomposite materials for specific applications. In this study, extraction of cellulose and nanocellulose from agricultural biomass (renewable resource) followed by preparation and characterisation of environmentally friendly polymeric materials and their composite products were studied. The study showed that the incorporation of nanocellulose into biopolymer matrix could produce bio-nanocomposites for specific uses in various applications, mainly in the biomedical and green packaging sectors.

Keywords: cellulose, nanocellulose, agricultural biomass, bio-nanocomposites, applications

INTRODUCTION

Landfill fees and costs are ever-increasing and the demand to recycle waste is increasing, particularly for developing countries like South Africa. Agricultural waste is one of the largest segments of the nationwide waste problems. These large volumes of agricultural waste threaten surface water and groundwater quality in the event of waste spills, leakage from waste storage facilities, and run-off from fields on which an excessive amount of waste has been applied as fertiliser. The objectives of recycling are to conserve resources and reduce the environmental impact of waste by reducing the amount of waste disposed of at landfills. The South African government in its Industrial Policy Action Plan (IPAP) strategy document identified the key industrial sectors critical to improving the country's gross domestic product (GDP) and associated job creation potential (DTI, 2013). Table 10.1 shows the profiles of the sectors in which some of the components could be replaced by bio-nanocomposites and biomaterials to comply with environmental legislations.

However, the new environmental regulations globally necessitated the need for research and development (R&D) of new eco-friendly materials as an alternative to non-biodegradable materials, particularly for conventional plastics from fossil fuel origin, which create serious environmental pollution and contributes to greenhouse gas emission (Salimon *et al.*, 2012). Large quantities of synthetic materials, such as glass, carbon, talc and aramid are widely used as fillers in polymer composites which have negative consequences as they pose end of useful life disposal problems (Chigondo *et al.*, 2013). In this respect, value-added, eco-friendly biomaterial products from agricultural biomass provide a new outlet for building bioeconomy and waste management strategies. The increase in the use of biobased materials and the proportionate decrease in the use of non-degradable synthetic materials could provide a better solution for growing environmental pollution (John & Thomas, 2008). The depletion of petroleum resources, increased oil price, greenhouse gas emission and overdependence on petroleum sources are the main driving forces for developing bioplastics and bioproducts from plant fibres. Currently, researchers are investigating the use of various plant-based fibres to develop sustainable eco-friendly materials as a viable substitute for synthetic reinforcements (Drzal *et al.*, 2001). The demand for plant fibres has drawn considerable attention to bio-nanocomposite applications due to their higher specific strength and stiffness as well as their eco-friendliness.

AGRICULTURAL BIOMASS

Maize stalks and sugar cane bagasse are grown in abundance in South Africa. Sugar cane is grown in 14 cane-producing areas extending from Eastern Cape Province through the coastal belt and KwaZulu-Natal midlands to the Mpumalanga Lowveld. Moreover, there are 22,500 registered sugar cane growers in South Africa that produce about 19 million tons of sugar cane from 14 million supply areas on average each season.[1] This has prompted increasing interest in the efficient use of agro-industrial by-products to reduce agricultural wastes produced every year, which poses potential pollution problems to the environment (Shaikh *et al.*, 2009). One of the by-products produced in large quantities by sugar and alcohol industries is sugar cane bagasse, which is commonly used for energy supply as a primary fuel source for sugar mills and also in the pulp and paper manufacturing sector (Chinnaraj & Rao, 2006; Brugnago *et al.*, 2011). However, the use of sugar cane bagasse in paper production requires the removal of short fibres, which leads to higher costs (Walford & Du Boil, 2006). On the other hand, the burning of sugar cane bagasse for energy supply is an inefficient means of utilising this potentially valuable residue, due to the low efficiencies of older boiler systems in most sugar mills.

Table 10.1: R&D focus and opportunities for composite and biocomposite materials for each identified sector (DTI, 2013)

IPAP focused sector	Aerospace	Automotive	Rail transport equipment	Renewable energy	Agro-processing	Plastics
Contribution to SA GDPs	*	R3.4bn (7% of GDP)	R44.2 billion (2.54% of GDP)	*	R7.7bn (16. % of GDP)	R50 bn (-R7bn trade deficit)
Sector objective	Substantially diversify and deepen the component supply chain.	Substantially diversify and deepen the component supply chain.	Metal fabrication, capital and rail transportation equipment.	Increase local content on renewable energy components.	Value addition of waste streams to increase beneficiation.	Address environmental concerns regarding plastic manufacturing and waste disposal.
Opportunities for composite and biocomposite materials	Increased use of composite and biocomposite materials for the manufacturing of high-performance, low cost, lightweight and reduced carbon footprint advantages.	Development of lightweight material automotive interior parts to reduce energy consumptions.	Development of lightweight composite and biocomposite material body panels with superior performance functionality (flame retardant, toughness, etc.).	Increase use of bioplastics production and manufacturing.	Extraction of high-value additives (biopolymers) and CNCs for food packaging and filtration applications.	Development of biodegradable polymers for use in plastic manufacturing.

*No authentic data available

Maize is abundantly produced in South Africa and is planted in Free State (34%), North West (32%), Mpumalanga (24%) and KwaZulu-Natal (3%) provinces annually in the rainy seasons.[2] Approximately 8 million tons of maize is produced yearly in South Africa on approximately 3.1 million hectares (ha) of land (Du Plessis, 2003). In developing countries like South Africa and other African countries, maize serves as a staple food for the majority of the population. The agricultural sector produces maize for the purpose of economic growth and after harvesting, maize stalk residues are left in the field as a "trash blanket" because they are not considered to have any direct use and are therefore underutilised (Du Plessis, 2003). Maize stalk residues are used in low-value products, such as animal feed or disposed of by incineration or dumped at waste sites due to the lack of space and little knowledge about possible value addition on such agricultural wastes.

Plant fibres are composed of cellulose as one of the major constituents, which is a principal structural component of plant cell walls. Other constituents of plant fibres are lignin, hemicellulose and extractives (Lu & Hsieh, 2012). Cellulose fibres are composed of bundles of microfibrils, of which each fibril – also called cellulose nanofibres (CNFs) – consists of crystalline and amorphous domains (Qing et al., 2013). CNFs provide a web-like structured network with high tensile strength and tensile modulus, high aspect ratios, large surface area and high crystallinity (Qing et al., 2013). The diameters and lengths of CNFs are in nano-scale (less than 100 nm) and micro-scale, respectively (Qing et al., 2013). The individualised nanofibres are obtained from cellulose by means of mechanical processes (Qing et al., 2013). The crystalline regions of nanofibres are called cellulose nanocrystals (CNCs), which are rod-shaped structures. CNCs remain after removal of amorphous domain either by sulphuric acid or enzyme hydrolysis (Satyamurthy et al., 2011; Silvério et al., 2013). In addition, the diameters and lengths of CNCs are in nano-scale (less than 100 nm) and micro-scale, respectively (Satyamurthy et al., 2011; Silvério et al., 2013). CNFs and CNCs are commonly termed as nanocellulose and cellulose is regarded as their main source. Thereafter, nanocellulose is used as reinforcements in the development of polymer composites. Recent research reports have indicated that nanocellulose offers a great potential in bio-nanocomposite applications as it provides comparative tensile strength and stiffness to synthetic reinforcements (Lu & Hsieh, 2012; Silvério et al., 2013; Lee et al., 2014). Other advantages of nanocellulose include inherent biodegradability and recyclability, renewable resource-based, low cost, low carbon footprint, environmentally benign nature and low density (Pullawan et al., 2010). With these advantages, the use of cellulose is becoming a popular solution in developing biobased products for various applications.

CELLULOSE AND NANOCELLULOSE EXTRACTIONS FROM BIOMASS

Agricultural fibres undergo a series of purification steps to eliminate lignin, hemicellulose, and extractives (pectin and waxes). These unwanted materials are mainly amorphous and they hinder extraction of pure cellulose and nanocellulose. The destruction of the cell wall after the treatment of fibres during a series of purification steps is shown in Figure 10.1.

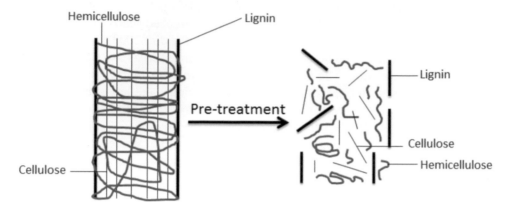

Figure 10.1: Treatment of lignocellulosic fibre [adapted from Salehudin *et al.* (2012)]

The purification steps are crucial for the extraction of high quality nanocellulose. Purification steps include chemical (alkali, organic solvent (dewaxing) and bleaching), enzyme hydrolysis and physicochemical treatments (steam explosion). The extraction of cellulose by removal of extractives, lignin and hemicellulose is shown schematically in Figure 10.2.

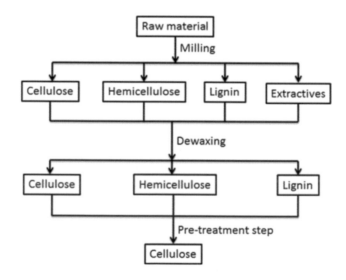

Figure 10.2: Schematic diagram of extraction of cellulose from plant fibres

Organic solvent treatment (dewaxing)

The main aim of this treatment is to remove extractives, such as waxes, fats, and oils from lignocellulosic fibres (Lu & Hsieh, 2012). This process is performed in a Soxhlet apparatus system using boiling organic solvent or a mixture of organic solvents over a certain period. The purpose of using solvents is to solubilise the solutes, in this case extractives (Jonoobi *et al.*, 2010; Maheswari *et al.*, 2012; Lu & Hsieh 2012; Mathew *et al.*, 2014). Factors that affect the selection of solvents are polarity of the solvent, boiling point, heat of vaporisation, dipole moment, density, viscosity and molar volume (Sin, 2012). Different organic solvents have been used for dissolving extractives. These solvents include mixtures of toluene and ethanol in a ratio of 2:1 (Lu & Hsieh, 2012), and acetone and methanol in a ratio of 2:1 (Jonoobi *et al.*, 2010). There are limited studies on the yield of fibres after solvent extraction. Mathew *et al.* (2014) studied the yield of fibres after solvent extraction. The purified solid matter obtained from the bioethanol plant in a Soxhlet extraction using a mixture of toluene/acetone (2:1 ratio) for six hours at 150°C. The yield of the fibres after the removal of extractives was found to be 85%.

Alkali treatment

The main aim of treating cellulosic fibres with alkali is to solubilise hemicellulose, lignin and extractives (Rosa *et al.*, 2010). Sodium hydroxide (NaOH) and potassium hydroxide (KOH) are commonly used alkalis. The treatment with alkali at higher temperatures of about 75°C to 80°C is used for solubilisation process (Motaung & Mtibe, 2015). A treatment with strong alkali could affect the crystallinity of the cellulose and degradation could occur (Ridzuan *et al.*, 2015). Following alkaline treatment, the material undergoes several repetitions of washes with water until a neutral pH level is achieved. This treatment swells the fibres and leaves some traces of alkali. These remaining traces of alkali make the fibres more accessible to other chemical treatments and mechanical processes.

Bleaching

Bleaching treatment delignifies cellulosic fibres. This process is performed by immersing the fibres in the bleaching solution at low pH levels (Lu & Hsieh, 2012). After this treatment, the colour of the resultant fibres changes to white (Silvério *et al.*, 2013). Different oxidising agents are used for delignification but chlorine dioxide (ClO_2) is the most popular (Hubbe *et al.*, 2008). Other common bleaching agents are hydrogen peroxide (H_2O_2), ozone, peracetic acid, and sodium hypochlorite (Hubbe *et al.*, 2008).

Steam explosion

This technique causes the breakdown of cellulosic fibres, either by defibrillation or by decompression. During the steam explosion process, the hot steam breaks the chemical bonds that co-exist between the various components of the complex structure of the cellulosic fibres (Singh *et al.*, 2014). Following heating by hot steam, the pressure is released to enhance defibrillation of cellulosic fibres. In addition, defibrillation of cellulosic fibres also paves the way for other treatments, for example, enzyme hydrolysis. Usually, in this technique the cellulosic fibres are heated using hot steam at temperatures varying between 160°C and 240°C and pressure varying between 0.7 MPa and 4.8 MPa (Singh *et al.*, 2014). Steam explosion is categorised as auto-hydrolysis and depressurisation, which hydrolyse hemicellulose into volatile compounds such as acetic acid, formic acid and furfural molecules. Acetic acid, which is produced from hemicelluloses hydrolysis acts as an auto-catalyst to further hydrolyse hemicellulose in sugars (Singh *et al.*, 2014). This can be achieved by opening the complex chemical structure of the fibres for treatment with enzymes. This technique can be used for hydrolysis of hemicellulose to break down the complex structure of cellulosic fibres to reduce their dimensions, to depolymerise lignin and for defibrillation of cellulosic fibres to produce cellulose nanofibres.

Enzyme hydrolysis

The treatment of biomass by microbial species, such as fungi, removes both ester and ether linked aromatics in lignin complex, producing high quality cellulose (Wan & Li, 2010; Wang *et al.*, 2013). This process also converts hemicellulose into sugars and other by-products. In addition, fungi loosen the complex structure of lignin and therefore degrade it (Wan & Li, 2010; Wang *et al.*, 2013). The enzyme hydrolysis process has received considerable interest because it is safe to use, environmentally friendly and no energy is required (Wan & Li, 2010; Wang *et al.*, 2013).

Characterisation of fibres before and after treatment, and the yield of final product

The characterisation of fibres before and after treatments can be performed to evaluate their properties through a range of characterisation techniques such as Fourier transform infrared (FTIR), thermogravimetric analysis (TGA), X-ray diffraction (XRD) and scanning electron microscopy (SEM). In addition, the qualitative analysis to investigate their chemical compositions is normally conducted using TAPPI standard methods.

The FTIR revealed that the removal of lignin and hemicellulose is characterised by the disappearance of C=O group, C-O and C=C at wave numbers of 1720–1750 cm^{-1}, 1242 cm^{-1} and 1512 cm^{-1} in treated fibres. These absorption bands represent the aliphatic carboxylic, aryl ester, and acetyl groups in the xylan component of hemicelluloses and lignin, deformation of the guaiacyl ring associated with the C-O stretch in lignin, and aromatic rings in lignin (Henrique *et al.*, 2013; Neto *et al.*, 2013; Motaung & Anandjiwala, 2015).

The surface morphologies of before and after treatment of natural fibres analysed by SEM reveal that the untreated fibres are normally bonded together by cementing materials (lignin and hemicellulose), however, after the treatment of fibres, the cementing material diminishes and the fibres bundles dispersed into individual fibres (Li *et al.*, 2009; Abraham *et al.*, 2011; Mtibe *et al.*, 2015a).

TGA results reveal thermal degradation behaviour of cellulose, hemicellulose and lignin at different temperatures. It is well reported that hemicellulose decomposes first, followed by cellulose and lignin respectively, with increasing temperatures (Silvério *et al.*, 2013). Motaung and Anandjiwala (2015) indicated that untreated fibres showed two decomposition stages, the decomposition of hemicellulose and α-cellulose, while the treated fibres showed one decomposition stage, the decomposition of cellulose, which confirmed that hemicellulose and lignin were removed in the treated fibres. The author results showed that the treatment of fibres with NaOH showed enhancement in thermal stability when compared to untreated fibres.

XRD analysis reveals the crystallinity behaviour of natural fibres. It was shown that the crystallinity of the fibres increases after the treatment of the fibres. This was attributed to the removal of non-cellulosic amorphous hemicellulose and lignin during the treatment (Mtibe *et al.*, 2015b).

Li *et al.* (2009) and Motaung & Anandjiwala (2015) indicated that the cellulose content increased after treatment with NaOH while hemicellulose and lignin were reduced. Costa *et al.* (2013) indicated that the total yield of pulp after treatment with soda was 38%. Li *et al.* (2012) discovered that the yield of pre-treated fibres was 72.7% after neutral sodium sulphite pre-treatment at 140°C with 7 wt.% alkali. The yield of bleached wood was found to be 71% to 77% (Mathew *et al.*, 2014).

PREPARATION OF NANOCELLULOSE

Cellulose nanofibres

Cellulose nanofibres are web-shaped materials, which consist of both crystalline and amorphous domains, extracted from the isolated cellulose by means of mechanical, combination of chemical and mechanical, and combination of enzymes and mechanical treatments. The dimensions of extracted CNFs are influenced by various factors, such as the source of cellulosic fibres, pre-treatment method and the mechanical technique applied. Yousefi *et al.* (2013) discussed the concept of extracting CNFs by means of repeated passes of the material during the mechanical processing, which causes the dimensions of the resultant material to decrease. Pre-treatment of fibres with enzymes and chemicals prior to mechanical treatment was found to play an essential role in the preparation of nanofibres from cellulosic fibres as these pre-treatments could weaken the bonds that link fibres.

Extraction of CNF's

Mechanical grinding

Micro-grinding is one of the preferred techniques due to low energy consumption during the extraction process and reduced clogging of the system (Yousefi *et al.*, 2013). The micro-grinding process involves passing the pulp slurry between two grinding stones, one stone remains static and the other rotates (Yousefi *et al.*, 2013). The mechanism of fibrillation in grinding is to break down hydrogen bonds and cell wall structure of plant fibres by shear forces and individualisation of pulp to nano-scale fibres (Yousefi *et al.*, 2013). The grinding stones have bursts and grooves that come into contact with the fibres to disintegrate them into the sub-structural components (Yousefi *et al.*, 2013). The material used to manufacture the grinding stones is usually non-porous resins containing silicon carbide. The supermass colloider (Model: MKCA6-3, Masuko Sangyo Co. Ltd, Japan), shown in Figure 10.3, is an example of a commercial micro-grinding system that mechanically fibrillates cellulosic fibres.

Figure 10.3: Supermass colloider (Model: MKCA6-3, Masuko Sangyo Co. Ltd, Japan)

Jonoobi *et al.* (2012) isolated nanofibres from cellulose fibres (CF) and sludge fibres (SF) where the researchers dispersed 3 wt.% of cellulose in water using mechanical blender, Silverson L4RT (England), at 3000 rpm for 10 minutes. Following that, the well-dispersed sample was ground in an ultra-fine grinder (Model: MKCA 6-3, Masuko Sangyo Co. Ltd Japan) to obtain nanofibers. The gap between the stones was adjusted and the process was performed in contact mode in motion (500 rpm). After feeding, the speed was increased to 1440 rpm or 3500 rpm, where the suspension was allowed to pass through a mechanical grinder until a gel was produced. In another study, Yousefi *et al.* (2013) allowed 1 wt.% water slurry as purified fibres to pass through a grinder (Model: MKCA6-3; Masuko Sangyo Co., Ltd., Japan) at a speed of 1500 rpm to produce CNFs.

High shear homogenisation

This process is generally carried out in combined steps of refining and high shear homogenisation (Saelee *et al.*, 2016). In this technique, cellulosic fibres are defibrillated by using high shear forces operated at pressures ranging from 100 to 1000 bars, which act on the fibre surface and at a temperature range from 60°C to 70°C (Besbes *et al.*, 2011). The homogenisation process is normally carried out at a high-speed range within 3500 rpm (Wong *et al.*, 2015) to 12 000 rpm (Chandra *et al.*, 2016). During the homogenisation process, the cellulose pulp is passed through a high shear pressure homogeniser for a number of times until translucent gel-like CNFs are produced (Besbes *et al.*, 2011).

Cellulose nanocrystals

Cellulose nanocrystals are highly crystalline rod-shaped materials obtained from the native fibres by acid and enzyme hydrolysis (Satyamurthy *et al.*, 2011; Silvério *et al.*, 2013). In the literature, different terms have been used to describe nanocrystalline cellulose, namely, nanocrystalline (Pirani & Hashaikeh, 2013), nanocrystals (Mtibe *et al.*, 2015a), and nanowhiskers (Mokhena & Luyt, 2014). To avoid confusion and inconsistency, the term cellulose nanocrystals (CNCs) was used throughout this section of the chapter.

The different shapes of CNCs could be attributed to their preparation methods. CNCs are highly crystalline materials with high aspect ratio (length to diameter ratio), large surface area and high strength and stiffness (Neto *et al.*, 2013). Their dimensions, crystallinity, stability and morphology are dependent on the type of cellulosic source and extraction method.

Extraction of CNCs

CNCs are extracted from cellulose by cleaving glycosidic bonds of amorphous domain of the cellulose chain (Silvério *et al.*, 2013; Neto *et al.*, 2013; Henrique *et al.*, 2013; Santos *et al.*, 2013), thus removing amorphous domains from the cellulose chain and leaving high, pure, individual rod-shaped unhydrolysed CNCs (Silvério *et al.*, 2013; Neto *et al.*, 2013; Henrique *et al.*, 2013; Santos *et al.*, 2013).

Acid hydrolysis

The literature survey demonstrates that acid hydrolysis is one of the preferred methods for extracting CNCs. Various acids, such as formic acid (Li *et al.*, 2015), hydrochloric acid (Henrique *et al.*, 2015) and sulphuric acid (Kallel *et al.*, 2016; Moriana *et al.*, 2016; Barana *et al.*, 2016) have been explored to extract CNCs. Among the acids used for hydrolysis, sulphuric acid (H_2SO_4) remains as the most popular and widely used. This could be due to the fact that H_2SO_4 forms negative sulphate groups on the surface of the CNCs through esterification (Neto *et al.*, 2013). Therefore, it increases the stability of the suspension containing CNCs thereby preventing flocculation of CNCs in the suspension (Rosa *et al.*, 2010). During acid hydrolysis, the hydronium ion penetrates the cellulose fibres, cleaves the glycosidic bonds of the cellulose fibrils and releases individualised CNCs (Neto *et al.*, 2013).

This process undergoes a series of steps. The first step is to immerse cellulosic fibres in strong acid under controlled time, temperature and acid-to-fibre ratio. The hydrolysis process is then terminated by adding tenfolds of water followed by the removal of excess acid using centrifugation in each step. The resulting suspension is normally dialysed against water for a couple of days to remove the excess acid until the suspension reaches pH levels between 6 and 7 (Rosa *et al.*, 2010). To prevent agglomeration this process is usually accompanied by dispersion technique (ultra-sonication) to obtain individualised cellulose crystals.

Enzyme hydrolysis

Similar to acid hydrolysis, enzyme hydrolysis is involved in the breaking of glycosidic bonds of amorphous domain, thereby producing individual crystals. The advantage of this method is its environmental friendliness. However, the major drawback of enzyme hydrolysis is that it produces less stable CNCs than the CNCs produced by acid hydrolysis (Satyamurthy *et al.*, 2011).

Various researchers have reported the comparison of enzymes and acid hydrolysis (Filson *et al.*, 2009; Satyamurthy *et al.*, 2011; Xu *et al.*, 2013). Satyamurthy *et al.* (2011) compared the effectiveness of acid and enzyme hydrolysis and reported that the CNCs obtained by H_2SO_4 hydrolysis showed better properties (crystallinity and dimensions) in comparison to those obtained by enzyme hydrolysis.

DEVELOPMENT OF BIO-NANOCOMPOSITES

Bio-nanocomposites are produced by reinforcing the biobased polymer matrix with nanocellulose. Polymer matrix provides the shape, structure and surface appearance, environmental tolerance, and overall durability of the resultant bio-nanocomposites. On the other hand, reinforcements are responsible for improving the properties of the resultant materials. Other functions of the polymer matrix are to hold the fibres together, transfer stress to those fibres and protect them from environmental and mechanical damage (Fowler *et al.*, 2006). The most widely used biobased polymers are polylactic acid (PLA) (Iwatake *et al.*, 2008), polyhydroxyalkanoates (PHA) (Ten *et al.*, 2010) and starch (Cao *et al.*, 2008). The composites produced from biobased reinforcement and biobased polymer – also called green composites, are likely to be more eco-friendly than the composites produced from synthetic polymers and biobased reinforcement.

Biobased polymers are produced from biobased natural resources and they have attracted considerable interest recently. They are synthesised from renewable plant materials or naturally occurring micro-organisms. Biopolymers could provide a perfect solution to producing sustainable and eco-friendly products, and they are regarded as promising polymers to replace petroleum-based synthetic polymers.

There has been a growing interest in the development of bio-nanocomposites for various applications, for example in the packaging and biomedical sectors (Fortunati *et al.*, 2012). Bio-nanocomposites are a relatively new and unique class of composites whereby reinforcement, having at least one dimension smaller than 100 nm, is employed.

The current research is aimed at finding a better combination of nanocellulose and polymer matrix, which will improve the properties and the performance of bio-nanocomposites. However, the bio-nanocomposite properties can be affected by natural fibre architecture (fibre geometry, dimensions, orientation, packing arrangement and volume fractions) and fibre-matrix interface (Fowler *et al.*, 2006). The geometry of the fibres is influenced by geographical conditions, whereas fibre extraction processes affect the properties of the bio-nanocomposites. The mechanical properties of the composites also depend on fibre volume fraction. The increase in volume fraction of the reinforcement leads to the improvement of mechanical properties until the maximum is attained (Fowler *et al.*, 2006). In addition, the dimensions of the fibres influence the mechanical properties of the resultant product. For example, fibres with smaller diameters and longer lengths provide high aspect ratio and large surface area, which result in the enhancement of interaction between the reinforcements and polymer matrices thereby improving the properties of the resultant material (Fowler *et al.*, 2006). At the interface between the reinforcements and polymer matrices the stress is transferred to the reinforcements via shear stresses thereby improving the mechanical properties (Fowler *et al.*, 2006).

In addition, the inclusion of nanocellulose in the biopolymer matrix exhibits extraordinary flexibility and physical properties of biocomposites. Furthermore, nanocellulose has fewer defects and therefore, they overcome the limitations of conventional micro-scale reinforcements. Also, due to the large surface area of nanocellulose, it presents a large volume of interface in composites with different properties in comparison to those of neat polymer matrix (Khalil *et al.*, 2014).

Cao *et al.* (2008) investigated the effect of variation in CNCs content from 5 to 20 parts per hundred parts of resin (phr) in plasticised starch bio-nanocomposites. The increase in CNCs content led to an increase in the tensile strength and tensile modulus but a decrease in elongation at break of the bio-nanocomposites produced from plasticised starch. Iwatake *et al.* (2008) produced PLA bio-nanocomposites reinforced with 5 wt.% of CNFs and observed the improvements in tensile strength and tensile modulus of the bio-nanocomposites in comparison to that of neat PLA. In the same study, the authors investigated the effect of CNFs content on the tensile properties of the bio-nanocomposites. They discovered that the increase in CNFs content from 3 to 20 wt.% led to an increase in both tensile strength and tensile modulus.

Chang *et al.* (2010) reported that the introduction of chitin nanoparticles in starch-based polymer results in an increase in the storage modulus of nanocomposites. However, the increase in nanoparticle content provided the improvement in stiffness. On the other hand, the damping factor of the nanocomposites shifted to the higher temperatures, which implied a good adhesion between the chitin nanoparticles and starch polymer.

Shi *et al.* (2012) produced PLA bio-nanocomposites reinforced with different contents of CNCs (1, 2, 5 and 10 wt.%). The increase in CNC content resulted in the improvement in thermal stability, except for PLA reinforced with 1 wt.% CNCs, which showed almost equal decomposition temperatures in comparison to that of the neat PLA. The PLA reinforced with 5 wt.% CNCs exhibited the highest thermal stability in comparison to other nanocomposites. This implied an improved heat resistance. The T_c of PLA initially decreased but also improved with the increase in CNC content. This was attributed to the increase in cold crystallisation at lower content while the aggregation occurred at high content in the bio-nanocomposites. It was also observed that at high content, two melt peaks were apparent due to the rearrangement of the PLA chains. The T_g of the PLA bio-nanocomposites reinforced with 5 wt.% of CNCs was higher in comparison to other bio-nanocomposites. The absolute crystallinity (X_p) and crystallinity (X_c) also increased with the increase in CNC content up to 5 wt.%. These results suggest that low reinforcement content can act as nucleation agents for improving crystallisation of PLA chains.

Cao *et al.* (2008) reinforced starch bio-nanocomposites with CNCs extracted from flax fibres. They observed small, white dots of CNC on the surface of nanocomposites.

These dots were less aggregated, thus indicating good dispersion and uniform distribution of the CNCs. According to this study, the uniform distribution was an indication of good adhesion between the CNCs and starch matrix. Similar observations were witnessed in thermoplastic starch polymer (TPS) reinforced with 10 wt.% CNFs. White dots of nanofibres were seen in the matrix and no fibre pullouts and debonding were observed. This was due to good adhesion between the CNFs and TPS, which provided good mechanical properties (Alemdar & Sain, 2008).

APPLICATIONS OF BIO-NANOCOMPOSITES

Nanocellulose is of great interest for a variety of applications, for example in the biomedical (Fernandes *et al.*, 2013) and food packaging sectors (Rhim *et al.*, 2013). However, earlier reports on extraction of cellulose nanocrystals by acid hydrolysis, have found that this hydrolysis process produces low yield of the product (CNCs) ranging from 30% to 44% (Bondeson *et al.*, 2006; Wang *et al.*, 2012), high production costs and also sometimes small aspect ratios (Dong *et al.*, 2016).

To achieve the broad market of nanocellulose, the yield of these materials must be improved. There has been ongoing research on scaling up the process of extracting CNCs, which enhances the efficiency and economy of nanocellulose and can contribute to CNCs to production cost reduction. The attempt to scale up the process was reported by Mathew and co-workers. In their study, the authors produced nanocellulose by integrating the laboratory scale with the pilot scale existing in wood-based industries. This process achieved the yield of CNCs in the range of 600 g/day (excluding the pilot-scale acid hydrolysis process and the cellulose-refining step). This yield was higher in comparison to 50 g/week in the laboratory scale process and the energy consumption of about 30 kWh (Mathew *et al.*, 2014).

The scientific community pursues research on developing biobased materials from renewable sources (reinforcement and polymer matrix) to provide good performance and improve their properties (Espino-Pérez *et al.*, 2013). Indeed, nanocellulosic reinforcements are suitable replacements for the traditional reinforcements due to their extraordinary properties. In addition, researchers are working to produce 100% biobased material or green materials for relevant food packaging and biomedical applications (Fortunati *et al.*, 2012).

Many other reports on developing films for food packaging have been documented. The all-cellulose nanocomposite films (ACNC) made from CNFs extracted from bagasse for food packaging were developed by Ghaderi *et al.* (2014). The TGA results showed that the thermal stability of ACNC was slightly lower than that of the films made from CNFs obtained from bagasse. The mechanical properties of ACNC were improved but the crystallinity was lower in comparison to the films made from CNFs. ACNC films exhibited acceptable values of water vapour permeability (WVP), which make them ideal for food packaging. Ninan *et al.* (2013) designed nanocomposites for scaffolds in tissue engineering. The novel nanocomposites for scaffolds were produced by combining pectin, carboxymethyl cellulose, and CNFs using the lyophilisation technique. The findings indicated that scaffolds were highly porous (10-250 µm) and bioactive. The addition of CNFs beyond the optimum content was found to reduce the thermo-mechanical and thermal properties of nanocomposites. The properties of the three dimensional scaffolds produced were suitable for tissue engineering, especially due to their high pore sizes.

Currently, nanocomposites are still at an early stage of commercialisation for many applications, as some improvements or modifications are required for their performance. These improvements include the increase of lifespan of these materials, flexibility, resistance to various weather conditions and cost. The price-performance ratio needs adjustments, especially when producing the fully biobased product. Currently, composites provide good mechanical properties but they lack consistency. Published research indicates that nanocomposites reinforced with nanocellulose perform better than biocomposites in terms of properties (Soykeabkaew *et al.*, 2012; Rhim *et al.*, 2013; Ghaderi *et al.*, 2014; Mathew *et al.*, 2014).

CONCLUSIONS

Nanocellulose as reinforcement in bio-nanocomposites for applications in the packaging and biomedical sectors has attracted considerable interest. There is an increasing demand for the development of biobased material due to the environmental problems caused by conventional materials. Cellulose and nanocellulose from sugar cane bagasse and maize stalks were chosen as waste biomass, which is abundantly available in South Africa. The results have indicated the potential for developing advanced composite materials. The cellulose nanocrystals (CNCs) are rod-shaped and their diameters and lengths were in nanoscale and micro-scale, respectively. Cellulose nanofibres (CNFs) are web-shaped and their diameters and lengths were also in nanoscale and micro-scale, respectively. The reinforcements of polymers with nanocellulose have shown improvement in the properties of the resultant composites and can be used in various applications such as green packaging, biomedical and automotive parts.

ACKNOWLEDGEMENTS

The authors would like to thank the Green Fund for the financial support of this project. The Green Fund is an environmental finance mechanism implemented by the Development Bank of Southern Africa (DBSA) on behalf of the Department of Environmental Affairs (DEA).

NOTES

1. http://www.sasa.org.za/sugar_industry/CaneGrowinginSA.aspx
2. http://www.nda.agric.za/docs/FactSheet/maize.htm

REFERENCES

Abraham, E, Deepa, B, Pothan, LA, Jacob, M, Thomas, S, Cvelbar, U, and Anandjiwala, R. 2011. Extraction of nanocellulose fibrils from lignocellulosic fibres: A novel approach. *Carbohydrate Polymers*, 86, 1468–1475.

Alemdar, A and Sain, M. 2008. Biocomposites from wheat straw nanofibers: Morphology, thermal and mechanical properties. *Composites Science and Technology*, 68, 557–565.

Barana, D, Salanti, A, Orlandi, M, Ali, DS and Zoia, L. 2016. Biorefinery process for the simultaneous recovery of lignin, hemicelluloses, cellulose nanocrystals and silica from rice husk and Arundo donax. *Industrial Crops and Products*, 86, 31–39.

Besbes, I, Vilar, MR, and Boufi, S. 2011. Nanofibrillated cellulose from Alfa, Eucalyptus and Pine fibres: Preparation, characteristics and reinforcing potential. *Carbohydrate Polymers*, 86, 1198–1206.

Bondeson, D, Mathew, A, and Oksman, K. 2006. Optimization of the isolation of nanocrystals from microcrystalline cellulose by acid hydrolysis. *Cellulose*, 13, 171–180.

Brugnago, RJ, Satyanarayana, KG, Wypych, F, and Ramos, LP. 2011. The effect of steam explosion on the production of sugarcane bagasse/polyester composites. *Composites Part A: Applied Science and Manufacturing*, 42, 364–370.

Chandra, JCS, George, N, and Narayanankutty, SK. 2016. Isolation and characterization of cellulose nanofibrils from arecanut husk fibre. *Carbohydrate Polymers*, 142, 158–166.

Cao, X, Chen, Y, Chang, PR, Muir, AD, and Falk, G. 2008. Starch-based nanocomposites reinforced with flax cellulose nanocrystals. *Express Polymer Letters*, 2, 502–510.

Chang, PR, Jian, R, Yu J., and Ma, X. 2010. Starch-based composites reinforced with novel chitin nanoparticles. *Carboydrate Polymers*, 80, 420–425.

Chigondo, F, Shoko, P, Nyamunda, BC, and Moyo, M. 2013. Maize stalk as reinforcement in natural rubber composites. *International Journal of Scientific & Technology Research*, 2, 263–271.

Chinnaraj, S, and Rao, GV. 2006. Implementation of an UASB anaerobic digester at bagasse-based pulp and paper industry. *Biomass and Bioenergy*, 30, 273–277.

Costa, SM, Mazzola, PG, Silva, JCAR, Pahl, R, Pessoa, A, and Costa, SA. 2013. Use of sugar cane straw as a source of cellulose for textile fiber production. *Industrial Crops and Products*, 42, 189–194.

DTI (Department of Trade and Industry). 2013. Industrial Policy Action Plan: 2013/14–2015/16.

Dong, S, Bortner, MJ, and Roman, M. 2016. Analysis of the sulfuric acid hydrolysis of wood pulp for cellulose nanocrystal production: A central composite design study. Industrial Crops and Products. http://dx.doi.org/10.1016/j.indcrop.2016.01.048. [Article in press.]

Dos Santos, RM, Neto, WPF, Silvério, HA, Martins, DF, Dantas, NO, and Pasquini, D. 2013. Cellulose nanocrystals from pineapple leaf, a new approach for the reuse of this agro-waste. *Industrial Crops and Products*, 50, 707–714.

Drzal, LT, Mohanty, AK, and Misra, M. 2001. Bio-composite materials as alternatives to petroleum-based composites for automotive applications. *Automotive Composites Conference*, 40, 1–8.

Du Plessis, J. 2003. Maize production. pp.1–38. Available at: http://www.cd3wd.com/CD3WD 40/LSTOCK/001/SA InfoPaks/docs/maizeproduction.pdf. [Accessed 10 August 2016].

Espino-Pérez, E, Bras, J, Ducruet, V, Guinault, A, Dufresne, A, and Domenek, S. 2013. Influence of chemical surface modification of cellulose nanowhiskers on thermal, mechanical, and barrier properties of polylactide based bionanocomposites. *European Polymer Journal*, 49, 3144–3154.

Fernandes, EM, Pires, RA, Mano, JF, and Reis, RL. 2013. Bionanocomposites from lignocellulosic resources: Properties, applications and future trends for their use in the biomedical field. *Progress in Polymer Science*, 38, 1415–1441.

Filson, PB, Dawson-Andoh, BE, and Schwegler-Berry, D. 2009. Enzymatic-mediated production of cellulose nanocrystals from recycled pulp. *Green Chemistry*, 11, 1808–1814.

Fortunati, E, Armentano, I, Zhou, Q, Iannoni, A, Saino, E, Visai, L, Berglund, LA, and Kenny, JM. 2012. Multifunctional bionanocomposite films of polylactic acid, cellulose nanocrystals and silver nanoparticles. *Carbohydrate Polymers*, 87, 1596–1605.

Fowler, PA, Hughes, JM, and Elias, RM. 2006. Biocomposites: technology, environmental credentials and market forces. *Journal of the Science of Food and Agriculture*, 87, 1132–1139.

Ghaderi, M, Mousavi, M, Yousefi, H, and Labbafi, M. 2014. All-cellulose nanocomposite film made from bagasse cellulose nanofibers for food packaging application. *Carbohydrate Polymers*, 104, 59–65.

Henrique, MA, Neto, FWP, Silvério, HA, Martins, DF, Gurgel, LVA, Barud, HS, Morais, LC, and Pasquini, D. 2015. Kinetic study of the thermal decomposition of cellulose nanocrystals with different polymorphs, cellulose I and II, extracted from different sources and using different types of acids. *Industrial Crops & Products*, 76, 128–140.

Henrique, MA, Silvério, HA, Neto, FWP, and Pasquini, D. 2013. Valorization of an agro-industrial waste, mango seed, by the extraction and characterization of its cellulose nanocrystals. *Journal of Environmental Management*, 121, 202–209.

http://www.nda.agric.za/docs/FactSheet/maize.htm. [Accessed 10 August 2016].

http://www.sasa.org.za/sugar_industry/CaneGrowinginSA.aspx. [Accessed 10 August 2016].

Hubbe, MA, Rojas, OJ, Lucia, LA, and Sain, M. 2008. Cellulosic nanocomposites: A review. *BioResources*, 3, 929–980.

Iwatake, A, Nogi, M, and Yano, H. 2008. Cellulose nanofiber-reinforced polylactic acid. *Composites Science and Technology*, 68, 2103–2106.

John, MJ, and Thomas, S. 2008. Biofibres and biocomposites. *Carbohydrate Polymers*, 71, 343–364.

Jonoobi, M, Harun, J, Mathew, AP, Hussein, MZB, and Oksman, K. 2010. Preparation of cellulose nanofibers with hydrophobic surface characteristics. *Cellulose*, 17, 299–307.

Jonoobi, M, Mathew, AP, and Oksman, K. 2012. Producing low-cost cellulose nanofiber from sludge as new source of raw materials. *Industrial Crops and Products*, 40, 232–238.

Kallel, F, Bettaieb, F, Khiari, R, García, A, Bras, J, and Chaabouni, SE, 2016. Isolation and structural characterization of cellulose nanocrystals extracted from garlic straw residues. *Industrial Crops and Products*, 87, 287–296.

Khalil, HPSA, Davoudpour, Y, Islam, MN, Mustapha, A, Sudesh, K. Dungani, R, and Jawaid, M. 2014. Production and modification of nanofibrillated cellulose using various mechanical processes: A review. *Carbohydrate Polymers*, 99, 649–665.

Lee, KY, Aitomäki, Y, Berglund, LA, Oksman, K, and Bismarck, A. 2014. On the use of nanocellulose as reinforcement in polymer matrix composites. *Composites Science and Technology*, 105, 15–27.

Li, B, Xu, W, Kronlund, D, Määttänen, A, Liu, J, Smått, JH, Peltonen, J, Willför, S, Mu, X, and Xu, C. 2015. Cellulose nanocrystals prepared via formic acid hydrolysis followed by TEMPO-mediated oxidation. *Carbohydrate Polymers*, 133, 605–612.

Li, Q, Gao, Y, Wang, H, Li, B, Liu, C, Yu, G, and Mu, X. 2012. Comparison of different alkali-based pretreatments of corn stover for improving enzymatic saccharification. *Bioresource Technology*, 125, 193–199.

Li, R, Fei, J, Cai, Y, Li, Y, Feng, J, and Yao, J. 2009. Cellulose whiskers extracted from mulberry: A novel biomass production. *Carbohydrate Polymers*, 76, 94–99.

Lu, P, and Hsieh, YL. 2012. Preparation and characterization of cellulose nanocrystals from rice straw. *Carbohydrate Polymers*, 87, 564–573.

Maheswari, UC, Reddy, KO, Muzenda, E, Guduri, BR, and Rajulu, AV. 2012. Extraction and characterization of cellulose microfibrils from agricultural residue- Cocos nucifera L. *Biomass and Bioenergy*, 46, 555–563.

Mathew, AP, Oksman, K, Karim, Z, Liu, P, Khan, SA, and Naseri, N. 2014. Process scale up and characterization of wood cellulose nanocrystals hydrolysed using bioethanol pilot plant. *Industrial Crops and Products*, 58, 212–219.

Mokhena, TC, and Luyt, AS. 2014. Investigation of polyethylene/sisal whiskers nanocomposites prepared under different conditions. *Polymer Composites*, 35, 2221–2233.

Moriana, R, Vilaplana, F, and Ek, M. 2016. Cellulose Nanocrystals from Forest Residues as Reinforcing Agents for Composites: A Study from Macro- to Nano-Dimensions. *Carbohydrate Polymers*, 139, 139–149.

Motaung, TE, and Anandjiwala, RD. 2015. Effect of alkali and acid treatment on thermal degradation kinetics of sugar cane bagasse. *Industrial Crops and Products*, 74, 472–477.

Motaung, TE, and Mtibe, A. 2015. Alkali treatment and cellulose nanowhiskers extracted from maize stalk residues. *Materials Sciences and Applications*, 6, 1022–1032.

Mtibe, A, Linganiso, LZ, Mathew, AP. Oksman, K, John, MJ, and Anandjiwala, RD. 2015b. A comparative study on properties of micro and nanopapers produced from cellulose and cellulose nanofibres. *Carbohydrate Polymers*, 118, 1–8.

Mtibe, A, Mandlevu, Y, Linganiso, LZ, and Anandjiwala, RD. 2015. Extraction of cellulose nanowhiskers from flax fibres and their reinforcing effect on poly furfuryl Alcohol. *Journal of Biobased Materials and Bioenergy*, 9, 309–317.

Neto, FWP, Silvério, HA, Dantas, NO, and Pasquini, D. 2013. Extraction and characterization of cellulose nanocrystals from agro-industrial residue - Soy hulls. *Industrial Crops and Products*, 42, 480–488.

Ninan, N, Muthiah, M, Park, IK, Elain, A, Thomas, S, and Grohens, Y. 2013. Pectin/carboxymethyl cellulose/ microfibrillated cellulose composite scaffolds for tissue engineering. *Carbohydrate Polymers*, 98, 877–885.

Pirani, S, and Hashaikeh, R. 2013. Nanocrystalline cellulose extraction process and utilization of the byproduct for biofuels production. *Carbohydrate Polymers*, 93, 357–363.

Pullawan, T, Wilkinson, AN, and Eichhorn, SJ.2010. Discrimination of matrix-fibre interactions in all-cellulose nanocomposites. *Composites Science and Technology*, 70, 2325–2330.

Qing, Y, Sabo, R, Zhu, JY, Agarwal, U, Cai, Z, and Wu, Y. 2013. A comparative study of cellulose nanofibrils disintegrated via multiple processing approaches. *Carbohydrate Polymers* 97, 226–234.

Rhim, JW, Park, HM, and Ha, CS. 2013. Bio-nanocomposites for food packa,ging applications. *Progress in Polymer Science*, 38, 1629–1652.

Ridzuan, MJM, Majid, MSA, Kanafiah, SNA, Afendi, M, and Nurinam, MBM. 2015. Effects of alkaline concentrations on the tensile properties of Napier grass fibre. *Applied Mechanics and Materials*, 786, 23–27.

Rosa, MF, Medeiros, ES, Malmonge, JA, Gregorski, KS, Wood, DF, Mattoso, LHC, Glenn, G, Orts, WJ, and Imam, SH. 2010. Cellulose nanowhiskers from coconut husk fibers: Effect of preparation conditions on their thermal and morphological behavior. *Carbohydrate Polymers*, 81, 83–92.

Saelee, K, Yingkamhaeng, N, Nimchua, T, and Sukyai, P. 2016. An environmentally friendly xylanase-assisted pretreatment for cellulose nanofibrils isolation from sugarcane bagasse by high-pressure homogenization. *Industrial Crops & Products*, 82, 149–160.

Salehudin, MH, Saleh, E, Muhamad II, and Mamat, SNH. 2012. Cellulose nanofiber isolation and its fabrication into bio-polymer. A review. *International Conference on Agricultural and Food Engineering for Life*, Cafei2012, 26–28 November, pp.1–26.

Salimon, J, Salih, N, and Yousif, E. 2012. Industrial development and applications of plant oils and their biobased oleochemicals. *Arabian Journal of Chemistry*, 5, 135–145.

Satyamurthy, P, Jain, P, Balasubramanya, RH, and Vigneshwaran, N. 2011. Preparation and characterization of cellulose nanowhiskers from cotton fibres by controlled microbial hydrolysis. *Carbohydrate Polymers*, 83, 122–129.

Shaikh, HM, Pandare, KV, Nair, G, and Varma, AJ. 2009. Utilization of sugarcane bagasse cellulose for producing cellulose acetates: Novel use of residual hemicellulose as plasticizer. *Carbohydrate Polymers,* 76, 23–29.

Shi, Q, Zhou, C, Yue, Y, Guo, W, Wu, Y, and Wu, Q. 2012. Mechanical properties and in vitro degradation of electrospun bio-nanocomposite mats from PLA and cellulose nanocrystals. *Carbohydrate Polymers*, 90, 301–308.

Silvério, HA, Neto, WPF, Dantas, NO, and Pasquini, D. 2013. Extraction and characterization of cellulose nanocrystals from corncob for application as reinforcing agent in nanocomposites. *Industrial Crops and Products*, 44, 427–436.

Sin, EHK. 2012. The extraction and fractionation of waxes from biomass. PhD Thesis pp.1–334.

Singh, R, Shukla, A, Tiwari, S, and Srivastava, M. 2014. A review on delignification of lignocellulosic biomass for enhancement of ethanol production potential. *Renewable and Sustainable Energy Reviews*, 32, 713–728.

Soykeabkaew, N, Laosat, N, Ngaokla, A, Yodsuwan, N, and Tunkasiri, T. 2012. Reinforcing potential of micro- and nano-sized fibers in the starch-based biocomposites. *Composites Science and Technology*, 72, 845–852.

Ten, E, Turtle, J, Bahr, D, Jiang, L, and Wolcott, M. 2010. Thermal and mechanical properties of poly3-hydroxybutyrate-co-3-hydroxyvalerate/cellulose nanowhiskers composites. *Polymer*, 51, 2652–2660.

Walford, SN, and Du Boil, PGM. 2006. A Survey of Value Addition in the Sugar Industry. Proceedings of the South African Technologists' Association, 80, 39–61.

Wan, C, and Li, Y. 2010. Microbial delignification of corn stover by Ceriporiopsis subvermispora for improving cellulose digestibility. *Enzyme and Microbial Technology*, 47, 31–36.

Wang, FQ, Xie, H., Chen, W, Wang, ET, Du, FG, and Song, AD. 2013. Biological pretreatment of corn stover with ligninolytic enzyme for high efficient enzymatic hydrolysis. *Bioresource Technology*, 144, 572–578.

Wang, QQ, Zhu, JY, Reiner, RS, Verrill, SP. Baxa, U, and McNeil, SE. 2012. Approaching zero cellulose loss in cellulose nanocrystal CNC production: Recovery and characterization of cellulosic solid residues CSR and CNC. *Cellulose*, 19, 2033–2047.

Wong, JCH, Kaymak, H, Tingaut, P, Brunner, S, and Koebel, MM. 2015. Mechanical and thermal properties of nanofibrillated cellulose reinforced silica aerogel composites. *Microporous and Mesoporous Materials*, 217, 150–158.

Xu, Y, Salmi, J, Kloser, E, Perrin, F, Grosse, S, Denault, J, and Lau, PCK. 2013. Feasibility of nanocrystalline cellulose production by endoglucanase treatment of natural bast fibers. *Industrial Crops and Products*, 51, 381–384.

Yousefi, H, Faezipour, M., Hedjazi, S, Mousavi, MM, Azusa, Y., and Heidari, AH. 2013. Comparative study of paper andnanopaper properties prepared from bacterial cellulose nanofibers and fibers/ground cellulose nanofibers of canola straw. *Industrial Crops and Products*, 43, 732–737.

11

VALORISATION OF MANGO SEEDS VIA EXTRACTION OF STARCH: USING RESPONSE SURFACE METHODOLOGY TO OPTIMISE THE EXTRACTION PROCESS

T Tesfaye[1, 2] and BB Sithole[1, 3]

[1] Discipline of Chemical Engineering, University of KwaZulu-Natal, Durban, 4041, South Africa

[2] Ethiopian Institute of Textile and Fashion Technology, Bahir Dar, Ethiopia

[3] Biorefinery Industry Development Facility - CSIR/UKZN, 35 King George Ave, Durban, 4041, South Africa

Corresponding author e-mail: BSithole@csir.co.za

ABSTRACT

Mango seed, a waste material that is disposed of after consumption of mangoes, was studied for its potential use as a resource for extraction of starch. The study revealed that mango seeds are a good source of starch. Physico-chemical characterisations confirmed that the extracted material was indeed a starch material. The starch was tested for use in textile applications and the results indicated that the material performed as well as a standard starch sample. The extraction of starch from mango seeds is facile and does not require sophisticated technology. Response surface methodology was used to optimise the extraction of starch from mango seeds. The experimental parameters optimised were concentration of mango seeds, extraction temperature, and extraction time, while the measured response factors were starch yield and whiteness of the starch. Thus, the critical values for optimal whiteness of the extracted starch were calculated to be: concentration = 0.35 (%w/v); temperature = 26.74°C; extraction time = 6.46 hours. A techno-economic analysis of the starch extraction process showed that the technology is viable and could be taken up by small-; medium- and micro-sized enterprises (SMMEs).

Keywords: mango, starch, extraction, response surface methodology, textile, sizing

INTRODUCTION

Starch, a glucose biopolymer, is the major storage component of most economically important crops, for example, cereals, legumes, tubers and yams (Yadav *et al.*, 2010; Musa *et al.*, 2011; Emmambux & Taylor, 2013; Alcázar-Alay & Meireles, 2015). It is a very versatile material with a wide range of applications in the food, pharmaceutical, textile, paper, cosmetic and construction industries (Jane, 1995; Ellis *et al.*, 1998;

Elliason, 2004; Tadesse *et al.*, 2015). Although South Africa is a large producer of starch, it nevertheless imports significant amounts of starch to meet its basic needs (IDC, 2016; International Starch Institute, 2016). However, increased importation and transportation costs, reduced availability, late deliveries and food security concerns are some of the major challenges facing the starch industries. Increased local supply of starch that does not compete with the food market is needed. Therefore, there is a need to investigate new botanical sources of starch.

Mango belongs to the botanical family *Anacardiaceae*. It is one of the most favoured commercially valuable fruits grown throughout the tropics and is used in a variety of food products (Yeshitela *et al.*, 2004). *Tommy Atkins*, *Sensation*, *Kent*, *Heidi*, *Keitt* and *Zill* varieties are the most common types of mangoes in South Africa. Approximately 84% are planted under micro, drip, sprinkler or flood irrigation systems (DAFF, 2014). Dryland production is no longer favoured, except where the annual rainfall supplements the irrigation programme during critical periods. Approximately 20% of mango producers account for 80% of the total annual production of 80,000 tons (National Agricultural Marketing Council, 2013). The weight of a mango seed accounts for about 15% to 25% of the total weight of the fruit. Therefore, each year about 16,000 tons of seeds are disposed of as a waste (Honja, 2014). However, the theoretical starch content of mango seed is 74% (Velan *et al.*, 1995); thus, mango seeds are a potential useful resource for starch. Some work has been done to valorise mango seeds. For example, Henrique and co-workers (2013) have done this by extraction and characterisation of its cellulose nanocrystals. However, the authors are not aware of valorisation of mango seeds via extraction of starch.

The objective of this study, as part of long-term research at the counsel for Scientific and Industrial Research (CSIR), is to avoid sending biowastes to landfills by converting them into valuable products instead. Thus, in this case the objective was to avoid the landfilling of mango seed waste by extraction of the starch content for industrial applications. Optimum conditions for extraction of starch from mango seeds were evaluated and the veracity of the extracted starch was confirmed by determination of physico-chemical properties of the extracted starch, compared to those of a commercial starch sample. After extraction, the starch was evaluated for use as a replacement for commercial starch used in the textile industry.

MATERIALS AND METHODOLOGY

Collection and preparation of mango seeds

The seeds of the *Tommy Atkins* mango variety were collected from the mango juice making industry.

Moisture content of the seeds: The mango seeds were washed in free-flowing water and the moisture content was determined using a method described by ISO (1997), which entails drying the seeds in an oven for 24 hours at 110°C (Umerie and Ezeuzo, 2000).

Starch extraction: The seeds were dried (3 hours at 105°C) and ground to a fine powder via pulverisation in a hammer mill and sieving through a 30-mesh standard sieve. The powder was steeped in a sodium metabisulphite solution (0.01% (w/v) at a known temperature (as per experimental design) and blended using a heavy-duty blender. The homogenate was sieved on 20 μm nylon mesh and washed with distilled water. The mixture was allowed to settle, after which the supernatant was discarded and the crude extracted starch was washed repeatedly with tap water until the wash water was clear. The starch extract was then dried (at 110°C) and stored at room temperature.

Experimental design and statistical optimisation of the extraction process: Response surface methodology based on the Box-Behnken design was used for optimisation of experimental conditions (Box &Wilson, 1951). Three independent variables, namely, extraction time, extraction temperature, and concentration of the sodium metabisulphite were selected for the study and the experiments were designed using Design-Expert software (9.0.5) and JMP 12 software. In total 15 experiments were executed to optimise the process parameters according to the design (Table 11.1). The response factors (starch yield and whiteness of the product) were determined by the coefficient of variation, analysis of variance and contour plots.

$$Y = \beta_0 + \beta_i X_i + \beta_{ij} X_i X_j + \beta_{ii} X^2_i \qquad (1)$$

Where Y is the predicted response variable, β_o, β_i, β_{ii}, β_{ij} are constant regression coefficients of the model, and X_i, X_j ($i=1, 3; j=1, 3; i{\neq}j$) represent the coded values of independent variables (Zhu, 2010).

Table 11.1: Coded and actual levels of the design factors

	Levels		
Independent factors	-1	0	1
A: Concentration (%w/v)	0.1	0.3	0.5
B: Temperature (˚C)	25	37.5	50
C: Time (Hr)	2	7	12

Statistical analysis

Analysis of variance (ANOVA): To determine the relative contribution of process conditions on percentage yield and whiteness index of the extracted starch, ANOVA was performed on the experimental data at 95% confidence level, which showed F observed versus F critical. The *p*-value was set at 0.05.

Response optimisation: With the help of Design-Expert software and JMP software the optimised responses were determined. In the software, the response optimiser searches for a combination of input variables that jointly optimise a set of responses by satisfying the requirements for each response in the set. The optimisation was accomplished by:

- Obtaining the individual desirability (d) for each response
- Combining the individual desirable to obtain the combined or composite desirability (D)
- Maximising the composite desirability and identifying the optimal input variable settings

To maximise the desirability of the parameters studied, the software employs a reduced gradient algorithm with multiple starting points that maximise the composite desirability to determine the numerical optimal solution.

Physico-chemical characterisation and analysis of starch extract

Moisture content: This was determined by drying starch samples for 3 hours at 105°C.
Ash content: This was ascertained by heating samples for 5 hours at 900°C.
Starch yield: The yield of the starch extraction was determined from the weight of the mango seed powder used and the final starch weight obtained from the procedure.
Functional group analysis: KBr discs of the starch samples were prepared and then measured for FTIR characterisations using a Nicolet Magna FR 760 FTIR spectrometer (Bruker). The spectra were recorded at room temperature using 64 scans at 2 cm^{-1} resolution from 400 to 4000 cm^{-1}.
pH: 5 g of starch extract in 20 ml distilled water was mixed thoroughly for 5 minutes, allowed to settle, and the pH of the water phase was measured.
Iodine test: 1 g of starch was boiled with 15 ml of water and allowed to cool. A few drops of 0.1N iodine solution were added to 1 ml of the mucilage and the colour change was recorded.
Water solubility index: Suspensions of 1 g of starch and 40 ml of distilled water were heated in a water bath at 50°C to 90°C for 5 minutes and 30 minutes. The suspensions were then cooled to room temperature and left to settle for 2 hours after which 10 ml was pipetted into a weighing dish and dried at 120°C for two hours to determine the soluble content. The remaining supernatant was carefully removed by suction and weighed to determine the water solubility index of the starch extract.
Foaming capacity: 2 g of starch sample was homogenised in 100 ml distilled water by using a vortex mixer for five minutes. The homogenate was poured into a 250 ml measuring cylinder and the volume occupied after 30 seconds was noted. The foaming capacity was expressed as a percentage increase in volume occupied by the starch solution. The mean of three replicate determinations was used.
Microscopic examination: Two drops of distilled water were placed on a clean slide, and 2 mg starch was dispersed in the water while ensuring that the starch grains settled down and were thinly spread on the slide. The slide was examined at different magnification up to 40X using a projection microscope. Twenty granules were randomly sampled for each treatment or variety and examined for size and shape.
Viscosity measurement: Viscosity was measured using a Ford viscosity cup. The starch solution was filled in the cup that was allowed to purge by allowing the viscous paste to exude through the orifice situated at the base of the cup. The time needed for purging the cup was measured carefully using a digital stopwatch at 3% and 5% concentration

of the starch solutions. The experiment was repeated three times and the average time of flow was calculated.

Use of the extracted starch in textile applications

Evaluation of industrial applications of the extracted starch was done by applying the starch for textile applications on cotton yarn. Important characteristics in textile applications include sizing, stiff finish, and ease of removal of the starch after application.

Sizing: The extracted starch was mixed with enough cold water to make a smooth, thin paste at a concentration of 3% and 5% W/W, and, using a glass rod, the mixture was stirred for 3 minutes. The prepared paste was then cooked for 30 minutes at 100°C. The sizing paste was applied to the cotton yarn, using a laboratory scale sizing machine: 10 cones of 20 count yarns were prepared in the creel and the machine was run at a speed of 20 m/min at a cylinder drying temperature of 150°C.

Strength regain and elongation at break: The samples were conditioned for 24 hours and the tensile properties of the yarn, before and after sizing, was measured using a single yarn strength tester according to ASTM D-2256 standard test method with 250 mm gauge length and 20±3 seconds breaking time. The experiment was repeated 10 times and the mean was calculated.

$$Strength\ regain\ (\%) = \frac{Strength\ after\ sizing - strength\ before\ sizing}{Strength\ after\ sizing} * 100 \qquad (2)$$

Hairiness: A Shirley hairiness meter was used to measure yarn hairiness, with an electronic sensor counting hairs that exceeded 3MM in a given length. The Sekisui procedure measures the hairiness level of unsized and sized yarn, reports the actual result, and calculates the percent reduction in hairiness.

Stiff finish: The extracted starch was mixed with enough cold water to make a smooth, thin paste at a concentration of 3% and 5% W/W and the mixture was stirred using a glass rod for 3 minutes. The solution was cooked and stirred until it thickened and the starch became transparent. The cotton fabric was saturated using a padder (one dip, one nip) to give 80% wet pick-up. The fabric was then dried at 100°C for 5 minutes in a hot-air oven. The stiffened fabric was ironed on both sides.

Bending length: Bending stiffness of fabrics was measured using a cantilever test at an inclination angle of 41.5° Bending rigidity properties of each group were investigated in warp and weft directions using the cantilever test method. The dimensions of tested specimens were 25x200 mm.

Ease of removal of the starch after application: From the stiffened treated fabric 10 cm×10 cm samples were prepared and weighed to 0.001 g accuracy. The samples were then boiled in water for 30 minutes after which the fabric was dried in a hot-air oven at 100°C for 3 minutes and weighed after cooling and conditioning of the fabric. The experiment was repeated three times and the average weight loss was calculated.

The aforementioned tests were also performed using a standard starch sample, maize starch, for comparison.

RESULTS AND DISCUSSIONS

The extraction of starch from mango seeds was facile and settling was not hampered by the presence of non-starch materials that remained suspended and floating and were easily decanted off. It is known that non-starch materials will not settle, due to the density difference of starch and other non-starch materials. For example, proteins are less dense than starch and would remain suspended at the top (Manek *et al.*, 2012). The extracted starch powder was off-white in colour and amorphous in nature. Physico-chemical properties of the extracted starch are shown in Table 11.2.

Table 11.2: Physico-chemical properties of mango seed starch

Sample	Moisture content of seed (%)	Moisture content of starch (%)	Ash content of starch (%)	Foaming capacity (%)	pH	Viscosity (sec)
Mango starch	44.4	5.68	1.54	51.23	5.78	30.4 (5% concentration) 13.8 (3% concentration)

Compared to a standard maize starch sample, the mango starch had the following characteristics:

* slightly lower pH;
* lower moisture content;
* lower ash content;
* higher foaming capacity;
* higher viscosity.

Moisture content: The moisture content of mango seeds is important as if affects the yield of the extracted starch. The moisture content of the extracted starch was below the recommended maximum of 14% (Table 11.2). This is desirable since low moisture content will not promote the growth of micro-organisms, like fungi, that will degrade the starch. This has implications for storage and shelf life of the extracted starch.

Iodine test and pH: The iodine exhibited a deep blue colour confirming that the extracted powder was starch. The pH of the starch extract ranged from 5 to 6, which was within the recommended range of between 4.5 and 7.0 (National Starch and Chemical Company, 2002).

Viscosity measurement: The viscosity at 5% concentration was too high to allow the starch to be used in a normal textile sizing application (Table 11.2). However, the viscosity at 3% concentration was good for such applications. This implies that the starch extract from mango seed can be used at lower concentrations, which has a positive impact from a cost point of view. Such properties are important for mixing and pumping operations in industries.

Microscopic examination: The particle shapes of the extracted starch were mainly oval to elliptical in shape (Table 11.3). This was in conformity with characteristics of standard starch. The large particle size is an advantage because large particles have smaller surface area and hence smaller surface activity. Particulate function is a surface phenomenon that generates resistance to flow. Thus, larger particles flow better than a smaller ones. The smaller particles (large surface area) have more surface energy to attract other particles and tend to adhere together creating more resistance to flow.

Table 11.3: Granule shape and size of mango seed starch

Sample	Granule size	Granule shape	
Mango	10.03 µm	Truncated, oval, round, elliptical	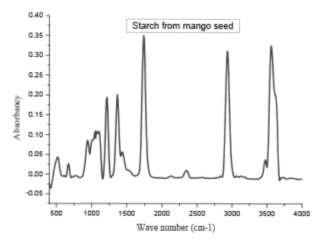

Functional group analysis: The FTIR spectrum in Figure 11.1 shows functional groups that are typical of starch materials: O–H stretching in the range 3700–3600 cm^{-1}; N–H stretching between 3400 and 3300 cm^{-1}; C–H bond around 2930 cm^{-1}; C–H aliphatic stretching the 3000–2850 cm^{-1} range; C–H bond adjacent to a double bond or aromatic ring; and C–H stretching wave number that increases and absorbs between 3100 and 3000 cm^{-1}; carbonyl stretching in the 1830–1650 cm^{-1} region; C=N stretching; -1, 4-glycosidic linkages (C–O–C) in the 930 and 1640 cm^{-1} region; C–O–H bonds in the 1080 cm^{-1} to 1158 cm^{-1} region; anhydro glucose ring between 990 cm^{-1} and 1030 cm^{-1} (Silverstein *et al.*, 2014).

Solubility profile: The results shown in Figures 11.2 and 11.3 indicate that starch solubility was affected by heating rate to a greater extent than swelling power was. Higher solubility values were obtained with increase in temperature. Nonetheless, at lower temperatures there were no major differences in the solubility behaviour of the extracted starch. However, differences started to develop when the temperature exceeded 75°C. This pattern is likely due to the chemical nature of amylose/amylopectin in the starch.

Figure 11.1: FTIR spectrum of extracted starch

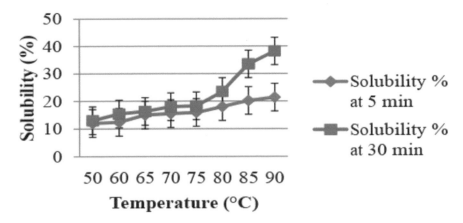

Figure 11.2: Effect of heating on solubility of starch at two heating times: 5 minutes (■) and 30 minutes (♦)

Swelling power: The swelling of starch granules was confirmed to be a two stage pattern: an initial low swelling was noticed up to 75°C and a big one thereafter (Figure 11.2). According to Figure 11.2, there were no statistical differences in swelling power for samples heated for 5 minutes and 30 minutes. However, differences were observed for samples heated above 75°C. It appears that the effect of heating time was significant at high temperatures due to extensive swelling of the granules. The swelling process was rapid during the first 5 to 10 minutes at the initial temperature and continued with further heating. At 50°C to 60°C amylose creates crystals with mango starch lipids, which inhibit excessive swelling of granules. At temperatures greater than 75°C the crystallites melt, and therefore swelling was enhanced, a fact that explains the fast swelling increase above 75°C.

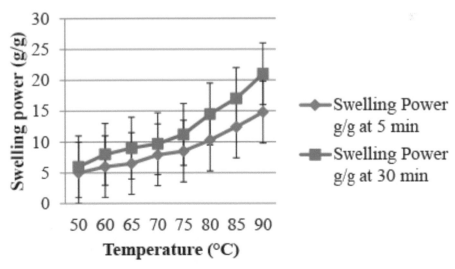

Figure 11.3: Effect of temperature on swelling power (g/g) of starch under two heating times, 5 minutes (■) and 30 minutes (♦)

Optimisation of starch extraction yield

The experimental parameters optimised were concentration of mango seeds, extraction temperature, and extraction time. The response factors that were measured were starch yield and whiteness of the starch. The actual values used are shown in Table 11.4.

Table 11.4: Testing conditions for starch extraction from mango seeds.

	Factor 1	Factor 2	Factor 3	Response 1	Response 2
Run	Concentration (%w/v)	Temperature (°C)	Time (hr.)	Yield (%)	Whiteness (%)
1	0.3	37.5	7	61.91	89.81
2	0.5	37.5	12	62.1	92.12
3	0.1	25	7	48.75	78.21
4	0.1	50	7	56.11	81.31
5	0.1	37.5	2	52.14	79.32
6	0.5	25	7	61.38	91.57
7	0.3	37.5	7	65.27	88.13
8	0.5	50	7	62.3	93.76
9	0.3	50	12	68.2	92.22
10	0.3	25	2	66.2	90.51
11	0.3	25	12	66.71	85.21
12	0.5	37.5	2	63.3	92.75
13	0.3	50	2	66.82	86.23
14	0.3	37.5	7	66.91	85.01
15	0.1	37.5	12	54.11	80.46

Fitting the model: The independent (concentration, temperature and time) and dependent variables (yield and whiteness) were analysed to obtain the regression equation of the model, which was an empirical relationship between the starch yield and the test variable in coded units, which could predict the response under the given range. The regression equation obtained for the extracted starch yield was as follows:

$$(Yield)^{1.69} = +402.61948 + 4987.39018 * Concentration - 4.5852 * Temperature - 20.71596 * time - 16.43025 * Concentration * Temperature - 21.71753 * Concentration * time + 0.1082 * Temperature * time - 5942.41491 * Concentration^2 + 0.15447 * Temperature^2 + 1.7883 * time^2$$

(3)

ANOVA (Tables 11.5 and 11.6) showed that this regression model was extremely significant ($p < 0.00001$). A model with p value less than 0.001 is highly significant. As shown in Tables 11.5 and 11.6, the F- and p-values of the lack of fit test were 46.52 and 0.000135, respectively, which implies that concentration of the mango seeds was the most significant variable. This indicates that the equation of the model was adequate for predicting the extraction of starch from mango seeds. The quadratic concentration F- and *p*-values of the lack of fit test were 65.90 and 0.000039, respectively, which implies that quadratic concentration was the most significant variable. The fitness of the model was further confirmed by a satisfactory value of the determination coefficient, which was calculated to be 0.6068, indicating that 60.68% of the variability in the response could be predicted by the model. The value of the adjusted determination coefficient (adjusted R^2 = 0.89242) also confirmed that the model was highly robust.

Table 11.5: The effect estimates of a full second-order polynomial model for optimisation of starch yield from mango seeds

Factors	Effect	Standard error	*t*-value	*p*-value	Coefficient	Standard error coefficient
Mean/interaction	60.68	0.57	106.79	0.000000	60.68	0.57
Concentration (%w/v)(L)	9.49	1.39	6.82	0.000135	4.75	0.70
Concentration (%w/v)(Q)	8.32	1.02	8.12	0.000039	4.16	0.51
Temperature (°C)(L)	2.59	1.39	1.87	0.098962	1.30	0.70
Temperature (°C)(Q)	-0.75	1.02	-0.74	0.482595	-0.38	0.51
Time (hr)(L)	0.67	1.39	0.48	0.645567	0.33	0.66
Time (hr)(Q)	-1.57	1.02	-1.49	0.173194	-0.77	0.51

R_2=0.93853; Adjusted R_2=0.89242 and Mean Square Residual=3.873993

Table 11.6: The ANOVA of a full second-order polynomial model for optimisation of starch yield from mango seeds

Factors	Sum of square	Degree of freedom	Mean square	F-test	*p*-value
Concentration (%w/v)(L)	180.2151	1	180.2151	46.51922	0.000135
Concentration (%w/v)(Q)	255.3345	1	255.3345	65.90990	0.000039
Temperature (°C)(L)	13.4940	1	13.4940	3.48323	0.098962
Temperature (°C)(Q)	2.1001	1	2.1001	0.54209	0.482595
Time (hr)(L)	0.8845	1	0.8845	0.22830	0.645567
Time (hr)(Q)	8.6622	1	8.6622	2.23598	0.173194
Concentration (%w/v) L+Q	435.5496	2	217.7748	56.21456	0.000019
Temperature (°C) L+Q	15.5941	2	7.7970	2.01266	0.195872
Time (hr) L+Q	9.5466	2	4.7733	1.23214	0.341603
Error	30.9919	8	3.8740		
Total sum of squares	504.1607	14			

The data for optimisation studies indicated that the effects of concentration on starch yield were statistically significant and there were significant relations of variables for starch yield as can be seen in Figure 11.4. The results demonstrated that yield was generally highest at a soak temperature of 50°C; soaking at 25°C resulted in a starch yield that was 3.15% to 5.25% lower. Starch yield was highest after a 12-hour steeping time but, unfortunately, this time length resulted in the development of microorganisms that were visible on the upper surface of the samples. Additionally, longer steeping times led to lower starch yields – this is mainly due to the increase in hydration and swelling of the starch, which reduces the filtrate amount making it difficult to facilitate sedimentation. The lower starch yield at low levels of sodium metabisulphite concentration and steep temperature was due to lower solubility and dispersibility of the starch molecule. The critical values for optimal whiteness of the extracted starch that were calculated by the software are as follows: concentration = 0.35 (%w/v); temperature = 26.74°C; time = 6.46 hours. The influential effect of the input variables on whiteness were evident using response surface plots as illustrated in Figure 11.4.

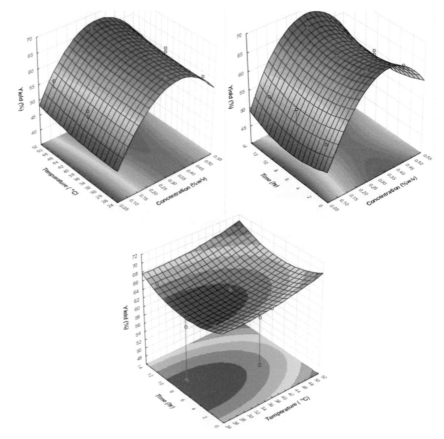

Figure 11.4: Response surface plot showing the effect of (concentration and temperature), (concentration and steeping time) and (temperature and steeping time) on yield of starch extraction from mango seeds *(a colour version is available from the authors)*

To further clarify the data and judge the adequacy of the model in the experimental data, diagnostic plots were drawn. A plot of observed response (yield) versus predicted response is shown in Figure 11.5 (A). In this case, the predicted values were in agreement with the observed ones in the range of the operating variables. The normal probability plot of the standardised residuals was used to check for normality of residuals (Figure 11.5 (B)). A linear pattern observed in this plot suggests that there were no signs of any problems in the experimental data. Figure 11.5 (C) represents a plot of standardised residuals versus predicted values to check for constant error. The residuals displayed randomness in scattering and suggested that the variance of the original observation was constant.

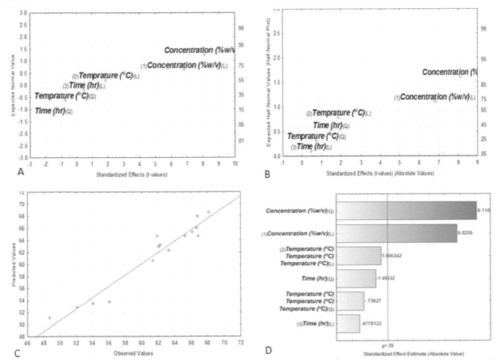

Figure 11.5: Normal probability design, half-normal probability design, observed vs. predicted plot and Pareto chart of standardised effects of starch yield respectively. (*Note: L-values are linear values of p-values, and Q-values are quadratic values of the variables. Q-values help to determine the optimal values of each of the variables and negative values are represented as positive values*)

Regression analysis (Table 11.5) and Pareto chart (Figure 11.5(D)) results indicate that concentration is the first and temperature is the second most influential variable among the chosen parameters as indicated by the p-values. It can be easily seen that the variable with the largest effect was the linear term of concentration, followed by quadratic concentration, and the linear of temperature. The factor t-test value (9.42) and p-value ($p = 0.000013$) corresponding to linear concentration and t-test value (8.11) and p-value ($p = 0.000039$) corresponding to quadratic concentration were the significant factors. According to the *t*- and *p*-values, temperature, time and quadratic values of both time and temperature did not exhibit statistical significance. The linear effect of time was found to have a *p*-value > 0.05 indicating the broad range effect of the variable on starch yield. Square values of the variables were used to ascertain their quadratic effects to determine the curvature in the response surface graphs and to determine the optimal value for each variable. The fit of the model was checked by determination of coefficient (R^2), which was 0.93853 thus revealing that 93.85% of the sample variation in starch yield was attributed to independent variables.

Optimisation of whiteness of the extracted starch

Fitting the models: The independent and dependent variables were analysed to derive a regression equation, which was an empirical relationship between the starch whiteness and the test variable in coded units that could predict the response under the given range. Independent and dependant variables were analysed to get regression equations that could predict the response under the given range; each of the observed value was compared with the predicted value, which was calculated from the model. The regression equation obtained for the extracted starch yield was as follows:

$$(Whiteness)^3 = +6.38561E + 005 + 7.870E + 005 * Concentration - 5553.35 * Temperature$$
$$- 36384.77093 * time - 275.29469 * Concentration * Temperature - 9494.2188$$
$$* Concentration * time + 1063.56025 * Temperature$$
$$* time \tag{4}$$

ANOVA (Tables 11.7 and 11.8) showed that this regression model was extremely significant ($p < 0.00001$). A model with *p*-value less than 0.001 is highly significant. The lack of fit test measures the failure of the model to represent the data in the experimental domain at points, which are not included in the regression. As shown in Tables 11.8 and 11.9, F- and *p*-values of the lack of fit test were 56.80 and 0.000067, respectively, which implies that concentration was the most significant factor, and indicates that the model equation was adequate for predicting the extraction of starch from mango seeds. The fitness of the model was further confirmed by a satisfactory value of the determination coefficient, which was calculated to be 0.8697, indicating that 86.97% of the variability in the response could be predicted by the model. The value of the adjusted determination coefficient (adjusted $R^2 = 0.7972$) also confirmed that the model was highly robust.

The optimisation studies indicated that the effects of concentration on whiteness of the extracted starch were statistically significant and there were significant relation of variables for whiteness. The whiteness index of the extracted starch was determined using 15 different recipes by varying the temperature, time and sodium metabisulphite concentration as shown in the experimental design in Table 11.4. The critical values for optimal whiteness percentage of the extracted starch given by the software are as follows: concentration (%w/v), 0.63; temperature, 24°C; and time, 6 hours. The influential effect of input variables on whiteness was represented using response surface plots as illustrated in Figure 11.6. Additionally, the results showed that the concentration of sodium metabisulphite could be reduced to 0.2% without loss of quality in starch purity. The whiteness degree of the extracted starch was higher when it was extracted in the presence of 0.5% concentration sodium metabisulphite and there was not much difference from the colour of a commercial starch, which is 95.75%.

Table 11.7: The effect estimates of a full second-order polynomial model for optimisation of starch whiteness from mango seeds

Factor	Effect	Standard error	t-value	p-value	Coefficient	Standard error coefficient
Mean/Interaction	86.97	0.69	126.18	0.000000	86.97	0.69
Concentration (%w/v)(L)	12.73	1.69	7.54	0.000067	6.36	0.84
Concentration (%w/v)(Q)	1.91	1.24	1.54	0.163071	0.95	0.62
Temperature (°C)(L)	2.01	1.69	1.19	0.269079	1.00	0.84
Temperature (°C)(Q)	0.47	1.24	0.38	0.714377	0.24	0.62
Time (hr)(L)	0.30	1.69	0.18	0.863387	0.15	0.84
Time (hr)(Q)	0.42	1.24	-0.34	0.743334	-0.21	0.62

$R^2=0.88411$; Adjusted R^2 =0.7972 and Mean Square Residual=5.701241

Table 11.8: The ANOVA of a full second-order polynomial model for optimisation of starch whiteness from mango seeds

Factor	Sum of square	Degree of freedom	Mean square	F-test	p-value
Concentration (%w/v)(L)	323.8513	1	323.8513	56.80365	0.000067
Concentration (%w/v)(Q)	13.4523	1	13.4523	2.35954	0.163071
Temperature (°C)(L)	8.0401	1	8.0401	1.41023	0.269079
Temperature (°C)(Q)	0.8200	1	0.8200	0.14382	0.714377
Time (hr)(L)	0.1800	1	0.1800	0.03157	0.863387
Time (hr)(Q)	0.6552	1	0.6552	0.11492	0.743334
Concentration (%w/v) L+Q	337.3035	2	168.6518	29.58159	0.000201
Temperature (°C) L+Q	8.8600	2	4.4300	0.77703	0.491597
Time (hr) L+Q	0.8352	2	0.4176	0.07325	0.929987
Error	45.6099	8	5.7012		
Total sum of squares	393.5772	14			

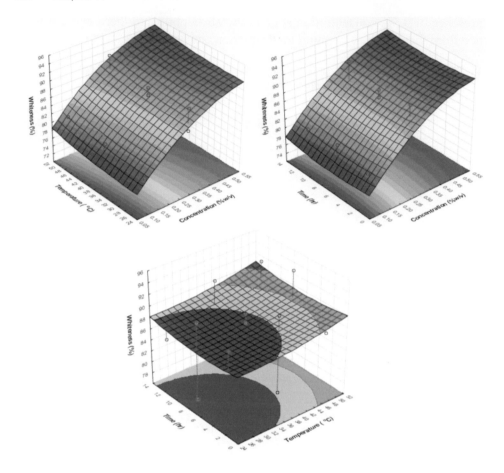

Figure 11.6: Response surface plot showing the effect of (concentration and temperature), concentration and steeping time) and (temperature and steeping time) on whiteness index of extracted starch respectively *(a colour version is available from the authors)*

To clarify the signs of any problems and judge the adequacy of the model in the experimental data, diagnostic plots were drawn. Plots of observed response (whiteness) versus predicted response are shown in Figure 11.7 (A) and (B). In this case, predicted values were in agreement with observed ones in the range of the operating variables. The normal probability plot of the studentised residuals was used to check for normality of residuals (Figure 11.8 (A) and (B)). A linear pattern observed in this plot suggests that there was no sign of any problem in the experimental data. Figure 11.7 (C) represents a plot of studentised residuals versus predicted values to check for constant error. Residuals displayed randomness in scattering and suggested that the variance of the original observation was constant.

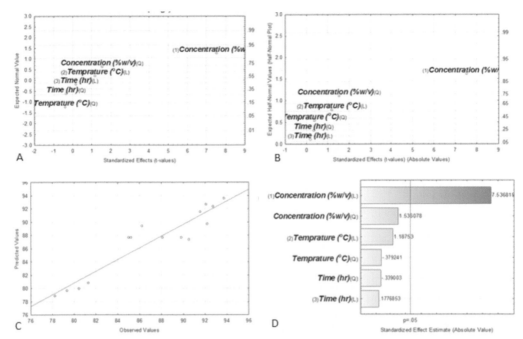

Figure 11.7: Normal probability design, half-normal probability design, observed versus predicted plot and Pareto chart of standardised effects of starch whiteness respectively

Regression analysis (Table 11.8) and Pareto chart (Figure 11.7 (D)) results indicate that concentration is the first and temperature the second most influencing variable among the chosen parameters as indicated by the p-values. The significance of each coefficient was determined using Pareto chart and p-value in Table 11.9 and it could be easily seen that the variable with the largest effect was the linear term of concentration, followed by quadratic concentration, and the linear of temperature. The factor t-test value (3.16275) and p-value ($p = 0.013340$) corresponding to linear concentration was the significant factor. According to the t-and p-value, temperature, time, the quadratic values of both time and temperature were of statistical significance. The linear effect of time was found to have a p-value > 0.05 indicating the broad range effect of the variable on starch whiteness. Square values of the variables are used to indicate their quadratic effects to determine the curvature in the response surface graphs and to determine the optimal value for each variable. The fit of the model was checked by determination of coefficient (R^2), which was 0.8841 revealing that 96.36% of the sample variation in starch yield was attributed to independent variables.

Testing of starch in textile sizing application

Strength regain and elongation: The strength regain of the sized yarn was within the recommended range of 15% to 40% (National Starch and Chemical Company, 2002). The major difference between the commercial starch extract and the extracted starch is that the commercial starch is a starch modified with additives that will have an impact on strength regain (Abbas *et al.*, 2010). These additives include arabic gum, defoamer and others. The size suspension from the extracted starch of the extracted starch was prepared without the additives. The elongation result was acceptable. It is important to note that it is generally expected for yarns to lose elongation after sizing operations due to stretching while the yarn is wet (Anonymous, 2016).

Table 11.9: Comparison of starches with respect to strength regain and elongation

	Untreated yarn		Yarn treated with mango seed starch		Yarn treated with commercial enset starch	
	Strength (N)	E %	Strength	E %	Strength	E %
Average	2.12	6.38	2.733	2.76	3.072	3.12
SR%			22.43		30.99	

SR = strength regain, E = elongation

Hairiness: It is important for the sizing material to coat the yarn surface well enough to slick down the "hairs" from the yarn bundle (Anonymous, 2016). The largest hairiness reduction of 61.2% was recorded for yarn sized with mango seed starch (Figure 11.8). The standard hairiness reduction of value is 50%. The greater the number of hairs, the greater the tendency to form a size bridge between ends on the slasher, leading to a harder break at the release rods, and the greater the amount of friction on the loom, resulting in excessive end breakage.

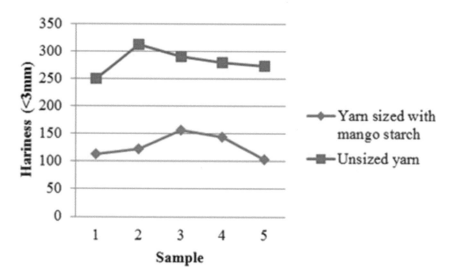

Figure 11.8: Hairiness value before and after sizing

Testing of starch in textile stiff finish applications

Bending length: For both 5% and 3% starch concentrations the bending lengths for fabric treated with the extracted starch and commercial starch were on the high side but there were no significant differences between the starch samples as shown in Table 11.10.

Table 11.10:   The bending property of the cotton fabric after and before stiff finish using extracted starch

Bending	Untreated fabric	Fabric treated with mango seed starch		Fabric treated with commercial enset starch	
		3% concentration	5% concentration	3% concentration	5 concentration
Average (cm)	2.4	5.9	8	5.6	8

Weight loss/ease of removal: A higher weight loss value indicates better removal of size mix from the treated fabrics. The data in Figure 11.11 indicate that mango seed starch is better for ease of removal of the size material from the fabric.

Table 11.11:   Comparison of starch removal characteristics during wet treatment of fabrics

Starch removal	Fabric treated with mango seed starch			Fabric treated with commercial enset starch		
Average	W1 (g)	W2 (g)	WL (%)	W1 (g)	W2 (g)	WL (%)
	2.57	1.61	27.35	2.12	1.4	33.96

W1 = weight of the fabric after stiffening treatment, W2 = weight of the stiffed finished fabric after washing with boiling water, and WL= weight loss (average of 3 replicates)

By-products of the starch extraction process

The by-products of the starch extraction process were characterised to ascertain whether they contained any valuable materials. The nitrogen content of the residue after extraction of the mango seeds was 0.73%. This value is comparable to the nitrogen content of an excellent natural fertiliser from sheep manure (0.9%) and greater than that exhibited by horse and cattle manure (0.5%). This implies that the residue of the starch extraction could be used as a good source of biofertiliser. Additionally, this by-product could be used as animal feed.

Techno-economic analysis of the starch extraction process

A techno-economic analysis of the starch extraction process from mango seeds was done in the Ethiopian context (Tesfaye & Sithole, 2017). The parameters and data that were considered in the analysis included:

- Establishment of a plant for the production of mango seed starch with a capacity of 500 tonnes per annum.
- The present demand for starch for use in five textile factories estimated at 1365 tonnes per annum.
- The starch demand is expected to reach 2225 tonnes by the year 2020.

A schematic representation of the starch extraction process is shown in Figure 11.9. A feasibility analysis of the starch process indicated that establishment of a starch extraction plant could create employment opportunities for 19 persons. The total investment requirement was estimated at about R3.05 million, of which R2.2 million would be required for infrastructure and machinery. The project was deemed financially viable with an accounting rate of return (ARR) of 83.28 % and a break-even analysis of 21%. Projections of this information to the South African context indicate that the extraction process could be viable option for creating SMMEs to establish and operate entities for extraction of starch from waste mango seeds.

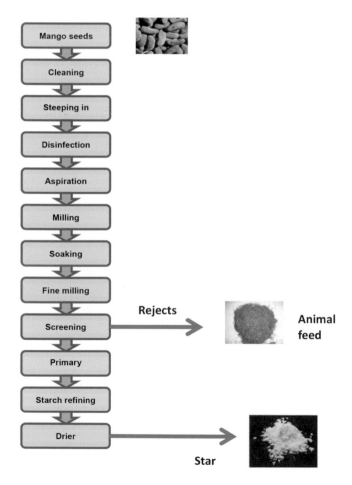

Figure 11.9: Process flow diagram for starch extraction from mango seeds

CONCLUSIONS

This study has revealed that mango seeds are a good source of starch: physico-chemical tests of the extracted starch confirmed that the extracted material was indeed a starch material. The starch was tested for use in textile applications and the results indicated that the material performed as well as a standard starch sample. Response surface methodology was used to optimise the extraction of starch from mango seeds. The experimental parameters optimised were concentration of mango seeds, extraction temperature, and extraction time and the response factors that were measured were starch yield and whiteness of the starch. Applications of the starch product in textile applications indicate that the material performs as well as a standard starch sample. The extraction of starch from mango seeds is facile and does not require sophisticated technology except that it is water-intensive.

An economic and financial feasibility analysis of the extraction of starch from mango seeds was done in the Ethiopian context and the results showed very good promise for the project. Thus, there is potential for the starch extraction process to be adopted by SMMEs in South Africa.

REFERENCES

Abbas, KA, Khalil, SK and Hussin, ASM. 2010. Modified starches and their usages in selected food products: A review study. *Journal of Agricultural Science*, 2(2), 90–100.

Alcázar-Alay, SC and Meireles, MAA. 2015. Physicochemical properties, modifications and applications of starches from different botanical sources. *Journal of Food Science and Technology (Campinas)*, 35 (2), 215–236.

Anonymous. 2016. Selvol™ Polyvinyl Alcohol for Textile Warp Sizing. Available at: http://www.sekisui-sc.com/wp-content/uploads/SelvolUltalux_WarpSizing_EN.pdf [Accessed 9 August 2016].

Bas, D and Boyaci, İH. 2007. Modeling and optimization I: Usability of response surface methodology. *Journal of Food Engineering*, 78(3), 836–845.

Box, G E P. and Wilson, KB. 1951. On the experimental attainment of optimum conditions (with discussion). *Journal of the Royal Statistical Society, Series B* 13(1), 1–45.

DAFF (Department of Agriculture, Forestry and Fisheries. 2014. A profile of the South African mango market value chain. Available at: http://www.nda.agric.za/doaDev/sideMenu/Marketing/Annual-Publications/Commodity-Profiles/FRUITS-AND-VEGETABLES/Mango-market-value-chain-profile-2014.pdf [Accessed 15 October 2015].

Eliasson, A-C. (ed.). 2004. *Starch in Food: Structure, Function and Applications*. CRC Press LLC, Boca Raton, FL.

Ellis, RP, Cochrane, MP, Dale, MFB, Duffus, CM, Lynn, A, Morrison, IM, Prentice, RDM, Swanston, JS and Tiller, SA. 1998. Starch production and industrial use. *Journal of the Science of Food and Agriculture*, 77(3), 289–311.

Emmambux, MN and Taylor, JRN. 2013. Morphology, physical, chemical, and functional properties of starches from cereals, legumes, and tubers cultivated in Africa: A review. *Starch*, 65(9–10), 715–729.

Henrique MA, Silvério, HA, Flauzino Neto, WP and Pasquini, D. 2013. Valorization of an agro-industrial waste, mango seed, by the extraction and characterization of its cellulose nanocrystals. *Journal of Environmental Manage*ment, 30(121), 202–209.

Honja, T. 2014. Review of mango value chain in Ethiopia. *Journal of Biology, Agriculture and Healthcare*, 4(25), 230–239.

IDC (Industrial Development Corporation). 2016. A study on the market potential for increased industrial starch. Available at: www.idc.co.za/images/2016/APCF-Starch-TOR.pdf [Accessed 6 August 2016].

International Starch Institute. 2016. Starch in South Africa. Available at: http://www.starch.dk/ISI/market/img/Starch%20in%20South%20Africa.pdf [Accessed 9 August 2016].

Jane, J. 1995. Starch properties, modifications, and applications. *J. Macromolecular Science, Part A: Pure and Applied Chemistry,* 32 (4), 751–757

Jayakody, L, Hoover, R, Liu, Q and Donner, E. 2007. Studies on tuber starches. II. Molecular structure, composition and physicochemical properties of yam (*Dioscorea sp.*) starches grown in Sri Lanka. *Carbohydrate Polymers,* 69(1), 148–163.

Ji, Y, Seetharaman, K and White, P. 2004. Optimizing a small-scale corn-starch extraction method for use in the laboratory. *Cereal Chemistry,* 81(1), 55–58.

Jobling, S. 2004. Improving starch for food and industrial applications. *Current Opinion in Plant Biology,* 7(2), 210–218.

Manek, RV, Builders, PF, Kolling, WM, Emeje, M and Kunle, OO. 2012. Physicochemical and binder properties of starch obtained from *Cyperus esculentus. Journal of the American Association of Pharmaceutical Sciences*, 13(2), 379–388.

Musa, H, Gambo, A, and Bhatia, P. 2011. Studies on some Physicochemical Properties of Native and Modified Starches from *Digitaria iburua* and *Zea mays. International Journal of the Pharmaceutical Sciences,* 3(1), 28–31.

National Agricultural Marketing Council. 2013. South African fruit trade flow. Available at: http://www.namc.co.za/upload/Fruit-Trade-Flow--June-2013-Issue-No-10.pdf [Accessed 15 October 2015].

Radosavljević , M, Jane, J and Johnson, L. 1998. Isolation of amaranth starch by diluted alkaline-protease treatment. *Cereal Chemistry,* 75(2), 212–216.

Shuren, J. 2000. Production and use of modified starch and starch derivatives in China *Cassava's Potential in Asia in the 21st Century: Present Situation and Future Research and Development Needs. Proceedings of a Sixth Regional Workshop, held in Ho Chi Minh city, Vietnam,* pp. 553–563.

Silverstein, RM, Webster, FX, Kiemle, D and Bryce, DL. 2014. *Spectrometric Identification of Organic Compounds.* 8th ed. New York: John Wiley & Sons.

Singh, N, Singh, J, Kaur, L, Sodhi, NS and Gill, BS. 2003. Morphological, thermal and rheological properties of starch from different botanical sources. *Food Chemistry,* 81(2), 219–231.

Tadesse, TF, Gebre, A, Gebremeske, AF and Henry, CJ. 2015. Pasting characteristics of starches in flours of chickpea (*Cicer arietinum* L.) and faba bean (*Vicia faba* L.) as affected by sorting and dehulling practices. *African Journal of Food Science*, 9(12), 555–559.

Tesfaye, T and Sithole, BB. 2017. Techno-economic feasibility of starch extraction from mango seeds. *Journal of Cleaner Production,* accepted — under revision.

Tian, S, Rickard, J and Blanshard, J. 1991. Physicochemical properties of sweet potato starch. *Journal of the Science of Food and Agriculture,* 57(4), 459–491.

Ubalua, A. 2007. Cassava wastes: treatment options and value addition alternatives. *African Journal of Biotechnology,* 6(18), 2065–2073.

Umerie, S and Ezeuzo, H. 2000. Physicochemical characterization and utilization of *Cyperus rotundus* starch. *Bioresource Technology,* 72(2), 193–196.

Vasanthan, T. 2001. Overview of laboratory isolation of starch from plant materials. *Current Protocols in Food Analytical Chemistry,* E:E2:E2.1, pp. 7–10.

Velan, M, Krishnan, M and Lakshmanan, C. 1995. Conversion of mango kernel starch to glucose syrups by enzymatic hydrolysis. *Bioprocess Engineering,* 12(6), 323–326.

Yadav, BS, Sharma, A and Yadav, RB. 2010. Resistant starch content of conventionally boiled and pressure-cooked cereals, legumes and tubers. *Journal of Food Science and Technology,* 47(1), 84–88.

Yeshitela, T, Robbertse, P and Stassen, P. 2004. Paclobutrazol suppressed vegetative growth and improved yield as well as fruit quality of 'Tommy Atkins' mango (*Mangifera indica*) in Ethiopia. *New Zealand Journal of Crop and Horticultural and Science,* 32(3), 281–293.

Zeng, M, Morris, CF, Batey, IL and Wrigley, CW. 1997. Sources of variation for starch gelatinization, pasting, and gelation properties in wheat. *Cereal Chemistry,* 74(1), 63–71.

THERMAL TREATMENT

EVALUATION OF DIFFERENT MUNICIPAL SOLID WASTE RECYCLING TARGETS IN SOUTH AFRICA IN TERMS OF ENERGY RECOVERY AND CO$_2$ REDUCTION

BO Oboirien[1] and BC North[2]

[1] Department of Chemical Engineering Technology, University of Johannesburg, 2028, South Africa

[2] CSIR Materials Science and Manufacturing, PO Box 395, Pretoria, 0001, South Africa

Corresponding author e-mail: boboirien@uj.ac.za

ABSTRACT

There is a move to divert waste from landfill by material recovery through waste recycling. There are two key targets in South Africa. PackagingSA (formerly PACSA) (2011) set a target to recycle paper by 61%, metal by 65%, glass by 43% and plastics by 36% by 2014. The second target is the reduction of garden waste to landfills in South Africa. The National Norms and Standards for Disposal of Waste to Landfill (RSA 2013) aims to divert 50% garden waste in 10 years (2023). The potential contribution of such targets on energy generation and carbon dioxide (CO$_2$) reduction needs to be evaluated and these have not yet been quantified. This study aimed to evaluate the potential amount of energy recovery and the quantity of CO$_2$ savings for the different targets. This was done by carrying out a detailed analysis of the potential energy and CO$_2$ savings from landfill biogas and incineration of municipal solid waste (MSW). Furthermore, 12 different scenarios that combine the recycling rate of both targets were modelled and the effect of the waste fraction was discussed. The results obtained showed that about 16,539 GWh/yr of electricity could be generated from waste incineration and 10,617 GWh/yr from landfill gas recovery if there was no recycling. The CO$_2$ emission savings was about 6.14 MtCO$_2$/yr for incineration and 10.74 MtCO$_2$/yr for landfill recovery. In terms of the PackagingSA recycling target, it was anticipated that a 21% reduction in the volume of MSW would be achieved and the amount of electricity from incineration would decrease by 33% from 16,539 GWh/yr to 11,128 GWh/yr. Landfill gas recovery was estimated to decrease by 17% from 10,617 to 8,803 GWh/yr. The CO$_2$ emission savings would be about 6.68 MtCO$_2$/yr for incineration and 8.93 MtCO$_2$/yr landfill gas recovery. The second target (50% recycling of garden) will lead to a 9% reduction in the volume of the MSW and the amount of electricity from incineration will decrease by 8% from 16,539 GWh/yr to 15,196 GWh/yr. Landfill gas recovery will decrease by 22% from 10,617GWh/yr to 8,323 GWh/yr. The CO$_2$ emission savings will be about 9.37 MtCO$_2$/yr for incineration and 8.44 MtCO$_2$/yr landfill recovery. Results from the analysis of the 12 different scenarios of both targets showed that the optimal condition for waste reduction and energy recovery will be at a recycling rate of paper 61%, metal 65%, glass 43%, plastic 27%, food 50% and garden waste 50%, and this will lead to a 41.8% reduction of the volume of the waste. The CO$_2$ savings will be about 5.87 MtCO$_2$/yr for incineration and 7.12 $MtCO_2$/yr for landfill gas recovery. The results also showed no correlation between the recycling rate and the amount of electricity that will be generated from both incineration and landfill gas. An analysis of the waste fractions showed that plastic and food waste have the highest impact on the amount of energy generated from incineration and garden waste on the amount of energy generated from landfill gas recovery.

Keywords: waste-to-energy, municipal solid waste (MSW), recycling targets, South Africa

INTRODUCTION

The management of municipal solid waste (MSW) is problematic worldwide. The severity of the problem is due to increased waste generation, as a result of increased economic development and population growth. The effective management of MSW could lead to pollution reduction and a sustainable source of energy (Corsten *et al.*, 2013; Charkraborty *et al.*, 2013). The Integrated Solid Waste Management (ISWM) process is the modern strategy for the effective management of MSW (UNEP, 2009). This involves the prevention of waste, followed by the reuse and recycling of waste, and finally the recovery, treatment and disposal of waste. There are three key methods of treating waste: thermal treatment, biological treatment and landfilling (Ofori-Baoteng *et al.*, 2013). The thermal treatment process involves incineration, gasification and pyrolysis while the biological process involves aerobic composting and anaerobic digestion (AD). The various treatment processes can also produce energy in addition to the reduction of the volume of the waste (Lino *et al.*, 2013). In South Africa, the Integrated Solid Waste Management (ISWM) process can also contribute to the social and economic development of its citizens, through better waste management systems, improved public health and the provision of jobs from recycling activities.

In South Africa, about 90% of the solid waste generated is disposed of in dumps and landfills (DEA, 2012). In 2011, approximately 59 million tonnes of general waste (including construction and demolition waste) were generated of which only 10% was recycled. In terms of municipal waste, an estimated 25.4 million tonnes are generated by households and disposed of to landfill (not recycled) (Friedrich & Trois, 2013). Recently, some local municipalities have been looking at recovering energy from waste because landfills are reaching capacity and there are limited new sites for waste disposal. The City of Johannesburg plans to generate 18 MW of electricity from landfill gas as part of the government's Renewable Energy Independent Power Producer Programme (REIPP, 2014). The gas will be collected from five landfill sites and the electricity generated will be fed into the national grid. The City of eThekiwini is also involved in similar projects (Couth & Trois, 2010). The landfill gas to electricity project is operated as a Clean Development Mechanism (CDM) project. In future, other municipalities could also start landfill gas collection and electricity generation (DEA, 2012).

There is also a move to divert waste from landfill to materials recovery (recycling) and energy recovery. Presently, two key targets support this intention. PackagingSA (formerly PACSA), which is a private initiative, set a five-year target to recycle paper by 61%, metal by 65%, glass by 43% and plastics by 36% by 2014 (PACSA, 2011). According to Godfrey (2016), some of these targets have been met – based on the actual recycling rate of 2015 (Paper 61.8%, metal 65.8%, glass 39.8% and plastics 34.8%). Future recycling targets for 2020 estimated by Friedrich and Trois (2016) are paper 71%, metal 75%, glass 53% and plastics 45%.

The second target is the reduction of garden waste in landfills in South Africa. The National Norms and Standards for Disposal of Waste to Landfill (RSA, 2013) (hereafter referred to as N&S) aim to divert 50% garden waste in 10 years (2023). Composting and other technologies could be used in the treatment of the garden waste to meet this target.

The potential contribution of such targets on energy generation and CO_2 reduction needs to be evaluated, as these have not yet been quantified. The aim of this study was to evaluate the potential amount of energy recovery and the quantity of CO_2 savings that could be realised by achieving the different targets. This was done by carrying out a detailed analysis of the potential energy and CO_2 savings from landfill biogas and incineration of MSW. Furthermore, 12 different scenarios that combine the recycling rate of both targets were modelled and the effect of the waste fraction is discussed.

METHODOLOGY

Physical and chemical composition of South African municipal solid waste (MSW)

The average physical composition of South African MSW reported by Friedrich and Trois (2013) was used in this study. The chemical composition of South African MSW has not been reported. The carbon content, moisture and calorific value (lower heating value – LHV) were taken from average values of municipal solid waste reported by Komilis et al. (2012). The fossil carbon ratio (FCR) was taken from Larsen and Astrup (2011), and the degradable organic carbon (DOC) fraction was taken from an International Panel on Climate Change report (IPCC, 2006). The values are presented in Table 12.1.

Table 12.1: Physical and chemical composition of the MSW

	Composition[a] (% w w)	Moisture content (%ww)	Carbon content (%dw)	FCR[b] (%)	DOC	LHV (MJ/kg)
Paper	18.2	6.1	39.9	0.5	40	14.45
Metal	3.9	5.3	0	0	0	0
Glass	6.9	2.0	0	0	0	0
Plastics	12	0.44	74.9	99.5	0	39.9
Food waste	26	71	48.0	1.1	15	4.3
Garden waste	18	41	43.2	2	63	9.2
Other	15	50	25	25	24	0

a - Average composition of South African MSW by (Friedrich and Trois, 2013)

Scenario analysis of both targets

The 2015 paper and packaging recycling rates and the 2020 targets are: paper (61.8% and 71%), metal (65.8% and 75%), glass (39.8% and 53%) and plastic (34.8% and 45%) (Friedrich & Trois, 2016). The 2011 recycling rate and the future recycling rate of organic waste are 35% and 50% respectively based on the National Waste Information Baseline Report (DEA, 2012) and the National Norms and Standards for Disposal of Waste to Landfill (RSA, 2013). Organic waste comprises food waste and garden waste.

In this study, we modelled the recycling rate of both targets. The sorting efficiencies were divided in two, with the low sorting efficiency being the actual recycling rate and the high sorting efficiency being the future recycling rate.

Twelve scenarios or combinations were derived and are presented in Table 12.2.

Table 12.2: Recycling scenarios (1–12)

	1	2	3	4	5	6	7	8	9	10	11	12
Paper (%)	71	71	71	71	71	71	62	62	62	62	62	62
Metal (%)	75	66	75	75	75	75	66	75	66	66	66	66
Glass (%)	53	53	40	53	53	53	40	40	53	40	40	40
Plastics (%)	45	45	45	35	45	45	35	35	35	35	35	35
Food (%)	50	50	50	50	35	50	35	35	35	35	50	35
Garden (%)	50	50	50	50	50	35	35	35	35	35	35	50

Calculation of energy potential and CO_2 emission from incineration and landfilling

Based on the calorific value of MSW for incineration and the total quantity of 25.4 million tonnes of household MSW that was not recycled in South Africa in 2011 (Friedrich & Trois, 2013), an overall efficiency of MSW energy to electricity of 23% was used (Tan et al., 2014). The emission factor (EF) was calculated by using the equation reported by Larsen and Astrup (2011). The equation was modified to report the emission factor (EF) in tonne CO_2/tonne of MSW rather than tonne CO_2/GJ and this is presented in Equation 1. Then, the total CO_2 emission was evaluated by multiplying the EF by the quantity of MSW incinerated which is presented in Equation 2.

$$EF= \frac{\sum_{i=1}^{n}(X_i*(1-SE_i)*TC_i*FCR_i)}{\sum_{i=1}^{n}(X_i*(1-SE_i))}a \tag{1}$$

EF=Fossil carbon emission factor of the MSW (tonne CO_2/tonne of MSW); Xi=share of waste fraction i in unsorted MSW [tonne wet/tonne wet]; SE_i=sorting efficiency of waste fraction i {tonne wet)/tonne (wet)}; FCR_i=fossil carbon ratio of waste fraction i}; a=conversion factor of 44/12.

$$\text{Total } CO_2 \text{ emission} = EF*\text{Quantity of MSW incinerated (Mt/y)} \qquad (2)$$

The potential energy that can be recovered from methane by the anaerobic decomposition of degradable organic carbon in the paper, food and garden waste fraction of the MSW was estimated by Equation 3 (Komilis *et al.*, 2013).

$$\text{CH}_4 \text{ generation from landfill} = \sum_i MSW * WF_i * MCF * DOC_i * DOCF * F * Y \qquad (3)$$

Where MSW= total waste not recycled that ends in landfill (tonnes). For this study, this was 25.4 million tonnes per year (Friedrich & Trois, 2013). The methane correction factor (MCF) is used to account for the fraction of waste that will decompose but will not produce methane. For this study, 0.8 was used as it was previously used in the estimation of emission of greenhouse gases (GHGs) from South African dumps by Friedrich and Trois (2013). DOC_p is the degradable organic carbon content. DOCF is the fraction of DOC dissimilated, and a value of 0.7 is recommended by Tsai (2007). F is the volumetric fraction of methane in the landfill gas and Y is the stoichiometric factor to convert C in CH_4 to CO_2. The global warming potential of CH_4 relative to CO_2 is 21(IPCC), hence the total CO_2 equivalent of CH_4 in terms of global warming, was obtained by multiplying by 21.

Estimation of CO_2 reduction

Although the recovery of energy from waste through incineration results in the emission of CO_2 from fossil carbon, there is an overall reduction of CO_2 because it will displace the consumption of coal for electricity generation in South Africa. The CO_2 emission savings are calculated in Equation 4.

$$CO_2 \text{ savings-}_{\text{(Incineration)}} = CO_2 \text{ emission from WtE} - \text{Amount of Electricity generated through WtE* CF} \qquad (4)$$

CF is the carbon avoided factor for every unit of power generation. In this study 1.015 $kgCO_2/kWh$ was used, as this is the average emission from Eskom coal power stations in South Africa (Letete *et al.*, 2011).

Since the CO_2 emission from landfill gas recovery is mainly of biogenic origin and not from fossil carbon, the CO_2 savings is equal to the amount of CO_2 displaced from the consumption of coal for electricity as shown in Equation 5.

$$CO_2 \text{ savings-}_{\text{(Landfill gas recovery)}} = \text{Amount of Electricity generated through WtE* CF} \qquad (5)$$

RESULTS AND DISCUSSION

Assessment of electricity generation potential and CO_2 reduction from MSW in South Africa

The potential of electricity generation and CO_2 reduction from MSW generated in South Africa was evaluated. In this analysis, electricity generated from MSW was assumed to be recovered through incineration or landfill gas recovery. The rate of electricity recovery from incineration was 0.65 MWh/tonne MSW and was 0.295 MWh/tonne MSW for landfill gas recovery (Figure 12.1). This indicates that energy recovery from the incineration of MSW is higher than in landfill gas recovery because of a higher energy recovery. Similar results were obtained by Tan *et al.* (2014) for Malaysian MSW. They obtained an energy recovery rate of 0.481 MWh /tonne MSW from incineration and 0.374 MWh/tonne MSW.

Based on the 25.4 million tonnes of MSW that were not recycled in South Africa in 2011 (Friedrich & Trois, 2013), the amount of electricity that could be generated from waste incineration is 16,539 GWh/yr and 10,617 GWh/yr could be generated from landfill gas recovery. The use of MSW as a substitute for fossil fuel for electricity generation can lead to CO_2 reduction (Tan *et al.*, 2014). The amount of fossil CO_2 emissions that would be avoided as a result of waste to energy (WtE) from incineration is 16.76 $MtCO_2$/yr and the amount of fossil CO_2 emissions from the combustion process itself would be 6.14 $MtCO_2$/yr.

The fossil emission factor for incineration was 0.37 $kgCO_2$/kWh. This is lower than the 1.015 kg CO_2/kWh from coal generated electricity in South Africa (Letete *et al.*, 2013), which would lead to a CO_2 emission saving of 10.67 $MtCO_2$/yr. Similar results were reported by Yang *et al.* (2012) for the greenhouse gas emissions from incineration of MSW from Ürümqi city in China, and Astrup *et al.* (2009) for greenhouse gas emissions of from incineration of MSW from European countries. Yang *et al.* (2012) reported a net emission rate of 26 $kgCO_2$/tonne MSW and Astrup *et al.* (2009) reported a net emission 47 to 966 kg CO_2/tonne MSW. In this study, we obtained a value of 412 $kgCO_2$/tonne MSW. For landfill recovery, the fossil emission factor will be 0, and the total avoided/saving CO_2 will be 10.74 $MtCO_2$/yr for landfill recovery. This value is close to the value of 10.67 $MtCO_2$/yr that was obtained for incineration.

Assessment of energy potential and CO_2 reduction of MSW in South Africa under future recycling targets

Based on the average composition of South African MSW given in Table 12.1, the five-year (2020) PackagingSA recycling target will lead to a 24.9% reduction in the volume of the waste. A WtE analysis was done on this target and the results are presented in Figure 12.2.

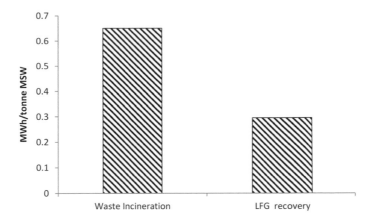

Figure 12.1: Assessment of potential electricity generation rate from MSW in South Africa

The rate of energy recovery from incineration will decrease from 0.62 MWh per tonne of MSW to 0.52 MWh per tonne of MSW and for landfill gas recovery it will increase from 0.295 MWh per tonne of MSW to 0.312 MWh per tonne of MSW. Although the volume of waste will reduce by 24.9%, the results showed that the amount of electricity from incineration will decrease by 33% from 16.539 GWh/yr to 13.333 GWh/yr and for landfill gas recovery it will increase by 8.4% from 10.617 to 11.511 GWh/yr. The CO_2 emission savings will be about 7.79 $MtCO_2$/yr for incineration compared to 10.67 $MtCO_2$/yr resulting in a 27% decrease in the CO_2 savings. For landfill gas recovery it will increase from 10.76 to 11.51 $MtCO_2$/yr resulting in a 7% increase in the CO_2 savings. This difference is due to a decrease in electricity generation rate for incineration and an increase in the energy rate for landfill gas recovery.

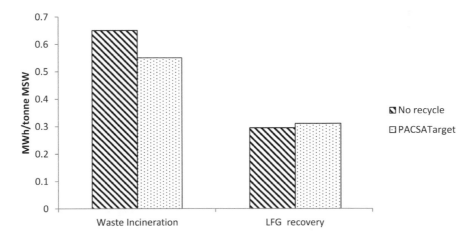

Figure 12.2: Assessment of the potential electricity generation rate from MSW in South Africa under PACSA targets

The second South African target is the reduction of garden waste disposed of in landfills in South Africa, according to the National Norms and Standards for the Disposal of Waste to Landfill (RSA, 2013) which aims to divert 50% in 10 years (2023). Based on the average composition of South African MSW, this would potentially lead to a 9% reduction in the volume of the MSW if other waste fractions are not recycled. A WtE analysis of this target showed that the rate of electricity recovery from incineration will increase slightly from 0.65 MWh per tonne of MSW to 0.66 MWh per tonne of MSW. For landfill recovery, it will decrease from 0.295 MWh per tonne of MSW to 0.255 MWh per tonne of MSW. The volume of waste will reduce by 9% and the amount of electricity from incineration will increase by 1% from 16.539 to 16.698 GWh/yr. There will be a 14% decrease in the amount of electricity from 10617 to 9135 GWh/yr for landfill gas recovery. The CO_2 emission savings will be about 10.31 $MtCO_2$/yr for incineration compared to 10.67 CO_2/yr resulting in a 3% decrease in the savings due to the increase in energy generation rate. For landfill gas recovery, it will decrease from 10.76 to 9.27 $MtCO_2$/yr resulting in a 14% decrease in the CO_2 savings.

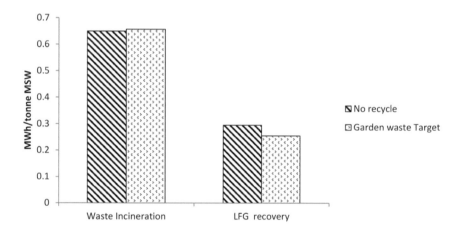

Figure 12.3: Assessment of the potential electricity generation rate from MSW in South Africa under National Norms and Standards (RSA, 2013) (Garden waste reduction)

Scenario analysis of both targets

A scenario analysis of both targets was carried out. Twelve different scenarios were used to model the possible realistic recycling rates based on the present and future recycling rates. Full details are presented in Table 12.2. The overall recycling rate and the individual waste fraction composition for each of the scenarios are presented in Table 12.3. The overall recycling rate was between 36.1% and 46.9%. An analysis of the potential electricity generation rate from incineration and landfill gas recovery is presented in Figure 12.4. The electricity generation rate for incineration ranges from 0.55–0.62 MWh/tonne MSW and 0.262–0.305 MWh/tonne MSW. The highest electricity generation rate from incineration was at a recycling rate of 45.7%. Landfill gas recovery was at 44.2%. The lowest energy generation rate from incineration was at a 43.0% recycling rate. Landfill gas recovery was at a 38.9% recycling rate.

This shows that there was no correlation between the recycling rate and the amount of electricity produced from both incineration and landfill gas. An analysis of the waste fractions showed that plastic and food waste have the highest impact on the amount of energy generated from incineration, and that garden waste has the highest input on the amount of electricity generated from landfill gas recovery. The recycling of plastic will lead to a reduction in the potential electricity that could be generated and the recycling of food waste could lead to an increase in the amount of potential electricity that would be generated.

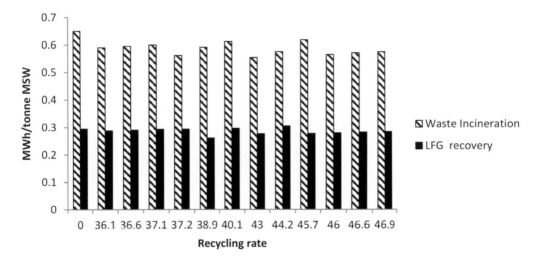

Figure 12.4: Energy assessment under 12 different scenarios

DISCUSSION

The ultimate aim of any waste management process is to optimally reduce the volume of waste that is landfilled, recover energy and reduce CO_2 emissions. In this section, we try to evaluate the optimal conditions that will result in the highest reduction of waste in landfill and highest energy recovery at the lowest emission. This was done by analysing the results of the two recycling targets and 12 other possible scenarios based on the two targets.

Table 12.3: Residual waste composition for the 12 scenarios (%ww)

Composition (%ww)	1	2	3	4	5	6	7	8	9	10	11	12
Paper	18.2	12.4	12.4	12.3	12.2	11.2	11.5	11.3	11.3	11.4	11.4	12.4	12.0
Metals	3.9	2.4	3.0	2.4	2.4	2.2	2.2	2.4	1.9	2.4	2.5	2.7	2.6
Glass	6.9	6.9	6.9	8.1	6.8	6.2	6.4	6.6	6.6	5.9	7.0	7.3	7.0
Plastics	12	13.5	13.4	13.3	15.1	12.1	12.5	12.3	12.4	12.4	11.0	13.5	13.1
Food waste	26	22.8	22.6	22.5	22.4	30.7	21.1	27.4	27.5	27.7	27.8	20.1	29.3
Garden Waste	18	15.8	15.7	15.6	15.5	14.2	21.9	19.0	19.1	19.2	19.3	20.9	13.5
Other	15	26.3	26.1	25.9	25.8	23.6	24.4	21.1	21.2	21.3	21.4	23.2	22.5
Total	**100**	**100**	**100**	**100**	**100**	**100**	**100**	**100**	**100**	**100**	**100**	**100**	**100**
Recycling rate	0	46.9	46.6	46.0	45.7	43.0	44.2	36.1	36.6	37.1	37.2	40.1	38.9

Volume reduction and energy generation rate

The potential amount of waste reduction from landfill, based on the two recycling targets and the 12 other possible scenarios, is presented in Table 12.4. The highest waste reduction will be in scenario 1 with a recycling rate of 46.9% and the lowest will be at the second recycling target at 9% recycling rate. However, the optimal condition for energy generation and volume reduction for incineration will be at scenario 4, with a recycling rate of 45.7% and energy generation rate of 0.618 MWh/tonne MSW. This is presented in Figure 12.5. For landfill gas recovery, the optimal condition for energy and volume reduction will be in scenario 6 with a recycling rate of 44.2%, and energy generation rate of 0.305 MWh/tonne MSW. This is presented in Figure 12.6.

Table 12.4:MSW volume reduction

Waste separation	Overall recycling rate
Target 1 - Paper 71%, Metal 75%, Glass 53%, Plastics 45%	24.9
Target 2 - Garden Waste 50%	9.0
Scenario 1 - Paper 71%, Metal 75%, Glass 53%,Plastics 45%, Food 50%, Garden 50%	46.9
Scenario 2 - Paper 71%, **Metal 66%,** Glass 53%, Plastics 45%, Food 50%, Garden 50%	46.6
Scenario 3 - Paper 71%, Metal 75%, **Glass 40%,** Plastics 45%, Food 50%, Garden 50%	46.0
Scenario 4 - Paper 71%, Metal 75%, Glass 53%, **Plastics 35%,** Food 50%, Garden 50%	45.7
Scenario 5 - Paper 71%, Metal 75%, Glass 53%, Plastics 45%, **Food 35%,** Garden 50%	43.0
Scenario 6 - Paper 71%, Metal 75%, Glass 53%, Plastics 6%, Food 50%, **Garden 35%**	44.2
Scenario 7 - Paper 62%, Metal 66%, Glass 40%, Plastics 35%, Food 35%, Garden 35%	36.1
Scenario 8 - Paper 62%, **Metal 75%,** Glass 40%, Plastics 35%, Food 35%, Garden 35%	36.6
Scenario 9 - Paper 62%, Metal 66%, **Glass 53%,** Plastics 35%, Food 35%, Garden 35%	37.1
Scenario10 - Paper 62%, Metal 66%, Glass 40%, **Plastics 45%,** Food 35%,Garden 35%	37.2
Scenario11 - Paper 62%, Metal 66%, Glass 40%, Plastics 35%, **Food 50%,** Garden 35%	40.1
Scenario12 - Paper 62%, Metal 66%, Glass 40%, Plastics 35%, Food 35%, **Garden 50%**	38.9

Volume reduction and CO_2 savings

The optimal conditions for volume reduction and CO_2 savings for the two recycling targets and the 12 other possible scenarios in incineration is scenario 11. The recycling rate is 40.1% and the CO_2 saving is 0.379 tonne CO_2/tonne MSW as presented in Figure 12.7. For landfill gas recovery, the optimal condition for energy and volume reduction will be scenario 6. The recycling rate is 44.2% and the CO_2 saving is 0.905 tonne CO_2/tonne MSW as presented in Figure 12.8.

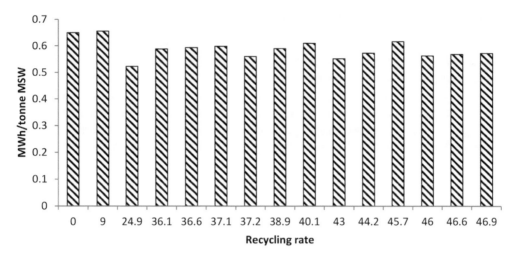

Figure 12.5: Comparison of electricity generation for incineration under the two recycling targets and 12 different scenarios

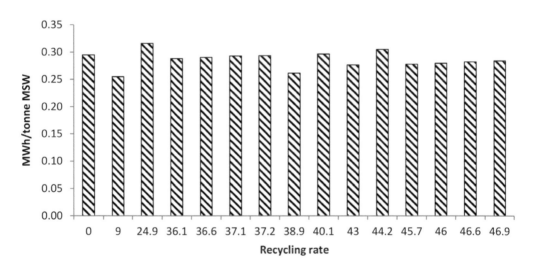

Figure 12.6: Comparison of electricity generation for landfill gas recovery under the two recycling targets and 12 different scenarios

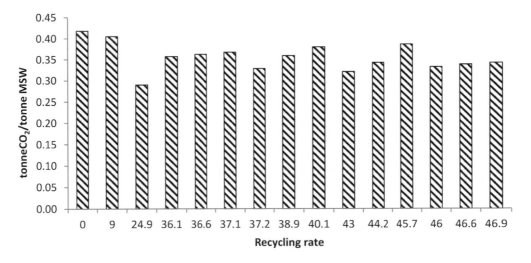

Figure 12.7: Comparison of CO_2 savings for incineration under the two recycling targets and 12 different scenarios

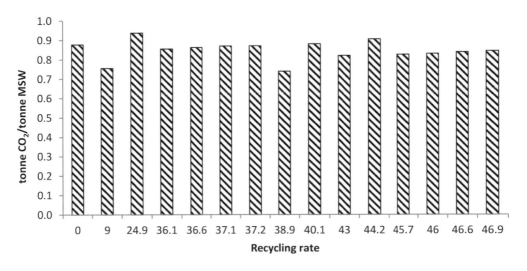

Figure 12.8: Comparison of CO_2 savings for landfill gas recovery under the two recycling targets and 12 different scenarios

Integration of incineration and landfill recovery

Results obtained so far have shown that there is a higher energy generation rate from waste incineration than from landfill gas recovery, however, landfill gas recovery results in a higher CO_2 saving due to the absence of fossil carbon emission. The CO_2 emission is mainly from biogenic sources such as food and garden waste. The integration of both strategies could assist in reducing the CO_2 emissions while maintaining a relatively high-energy generation rate. Based on the two recycling targets and the 12 possible scenarios investigated in this study, the conditions that give the optimal waste reduction and energy generation rate and CO_2 savings is scenario 11. Under this condition, the recycling rate of paper is 62%, metal 66%, glass 40%, plastics 35%, food 50% and garden waste 35%. The overall recycling rate is 40.1% and resulted in a mass reduction of 10.2 million tonnes.

CONCLUSIONS

This study has quantified the effect of different waste recycling targets and electricity generation from MSW in South Africa. The electricity generation and carbon reduction potential of two key recycling targets and 12 possible scenarios associated with them was analysed. The results also showed that no clear correlation between the recycling rate and the amount of electricity that will be generated from incineration and landfill gas. An analysis of the waste fractions showed that plastic and food waste have the highest impact on the amount of energy generated from incineration, and that garden waste has the highest impact on the amount of electricity generated from landfill gas recovery. The optimal condition for waste reduction, electricity generation and CO_2 savings will be at a recycling rate of paper 62%, metal 66%, glass 40%, plastics 40%, food 50% and garden waste 35%.

ACKNOWLEDGEMENTS

The authors gratefully acknowledge financial and other support received for this research from the CSIR South Africa

REFERENCES

Astrup, T, Moller, J and Fruergaard, T. 2009. Incineration and co-combustion of waste: accounting of greenhouse gases and global warming contributions. *Waste Management Research* 27: 789–799.

Charkraborty, M., Sharma, C., Pandey, J. and Gupta, P. 2013. Assessment of energy generation potentials of MSW in Delhi under different technological options. *Energy Conversion and Management,* 75, 249–255.

Corsten, M, Worrell, E, Rouw, M and Van Duin, A. 2013. The potential contribution of sustainable waste management to energy use and greenhouse gas emission reduction in the Netherlands. *Resources, Conservation and Recycling* 77,13–21.

Couth, B and Trois, C. 2010. Carbon emissions reduction strategies in Africa from improved Waste management – a review. *Waste Management* 30, 2347–2353.

DEA (Department of Environmental Affairs). 2012a. Municipal Waste Strategy Final. Available at: http:// sawic.environment.gov.za/documents/1352.pdf [Accessed June 2015].

DEA (Department of Environmental Affairs) 2012b. National Waste Information Baseline Report, Draft 6. Available at: sawic.environment.gov.za/documents/1567.docx. [Accessed June 2015].

Friedrich, E and Trois, C. 2013. GHG emission factors developed for the collection, transport and landfilling of municipal waste in South African municipalities. *Waste Management,* 32, 1013–1026.

Friedrich, E and Trois, C. 2016. Current and future greenhouse gas (GHG) emissions from the management of municipal solid waste in the eThekwini Municipality, South Africa. *Journal of Cleaner Production,* 112, 4071–4083

Komilis, D, Evangelou, A, Giannakis, G and Lymperis, C. 2012. Revisiting the elemental com position and calorific value of the organic fraction of municipal solid wastes. *Waste Management* 32: 372–381.

Larsen, AW and Astrup, T. 2011. CO_2 emission factors for waste incineration: Influence from source separation of recyclable materials. *Waste Management,* 31, 1597–1605.

Letete, T, Guma, M and Marquard, A. 2011. Carbon accounting for South Africa. Available at: http://www.erc.uct.ac.za/Information/Climate%20change/Climate_change_info3–Carbon_accounting.pdf [Accessed June 2015].

Lino, FAM, Ismail, KAR and Cosso, IL. 2013. Evaluation of the potential of recycling for the reduction of energy and CO_2 emissions in Brazil. *Sustainable Cities and Society*, 8, 24–30.

IPCC (International Panel on Climate Change). 2006. Guidelines for Greenhouse Gas Inventories, volume 5, Waste and volume 4, Agriculture, Forestry and other land uses.

Ofori-Boateng, C, Lee, KT and Mensah, M. 2013. The prospects of electricity generation from municipal solid waste (MSW) in Ghana: A better waste management option. *Fuel Processing Technology,* 110, 94–102.

PACSA (Packaging Council of South Africa). 2011. Packaging and Paper Industry Waste Management Plan. Submitted to the Department of Environmental Affairs, South Africa, Pretoria. 2011. Available at: http://www.pasca.co.za/pictures/pdfs/IndWMP.pdf (Accessed June 2015).

REIPP. 2014. Waste to Energy Project-REIPP programme. *Energy Journal,* 7, Available at: http://www.cefgroup.co.za/the-energy-journal [Accessed June 2015].

RSA (Republic of South Africa). 2013. National Norms and Standards for Disposal of Waste to Landfill. Available at: https://www.environment.gov.za/sites/default/files/gazetted_notices/nemwa59of2008_norms_standards_fordisposa.pdf [Accessed June 2016].

Tan, ST, Hashim, H, Lim, JS, Ho, WS, Lee, CT and Yan, J. 2014. Energy and emissions benefit of renewable derived from municipal solid waste: Analysis of a low carbon scenario in Malaysia. *Applied Energy,* 136, 797–804.

Tsai, WT. 2007. Bioenergy from landfill gas (LFG) in Taiwan. *Renewable and Sustainable Energy Reviews,* 11, 331–344.

UNEP (United Nations Developement Programme). 2009. Developing integrated solid waste management plan training manual. Available at: http://www.unep.org/ietc/Portals/136/Publications/Waste%20Management/ISWMPlan_Vol3.pdf [Accessed June 2016].

Yang, N, Zhang, H, Chan, M, Shao, L and He, P. 2012. Greenhouse gas emissions from MSW incincration in China: Impacts of waste characteristics and energy recovery. *Waste Management*, 32, 2552–2560.

INDEX